UNBELIEVABLE
SCIENCE

STERLING
New York

An Imprint of Sterling Publishing Co., Inc.
1166 Avenue of the Americas
New York, NY 10036

ISBN 978-1-4549-3123-2

Distributed in Canada by Sterling Publishing Co., Inc.
c/o Canadian Manda Group, 664 Annette Street
Toronto, Ontario M6S 2C8, Canada

For information about custom editions, special sales, and premium and
corporate purchases, please contact Sterling Special Sales at 800-805-5489 or
specialsales@sterlingpublishing.com.

Manufactured in China

2 4 6 8 10 9 7 5 3 1

sterlingpublishing.com

Picture Credits - see page 224

Introduction image: A human nerve cell
(neuron), derived from a stem cell.

UNBELIEVABLE
SCIENCE

STUFF THAT WILL
BLW
YOUR MIND

COLIN BARRAS

STERLING
New York

CONTENTS

CHAPTER ONE
SPACE

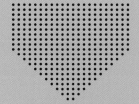

CHAPTER TWO
PHYSICS

CHAPTER THREE
TECHNOLOGY

CHAPTER FOUR
ENVIRONMENT

CHAPTER SEVEN
THE BRAIN
& HUMAN BEHAVIOR

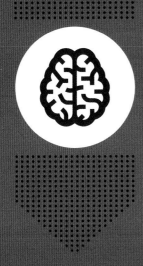

CHAPTER EIGHT
HUMANITY'S PAST, PRESENT & FUTURE

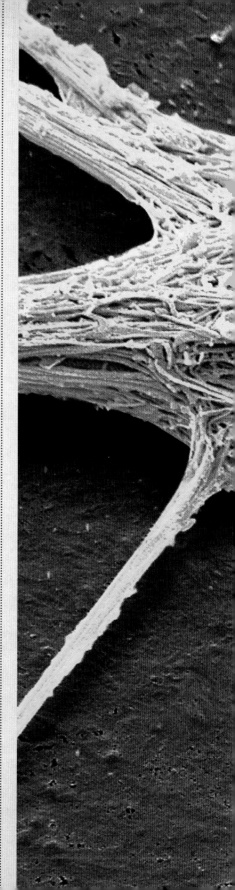

INTRODUCTION

Did you hear about the 430,000-year-old murder? Or the genes that whirr into gear days after death? How about the animals apparently thriving around the site of the world's largest nuclear disaster?

Some of the breakthroughs and discoveries made by scientists are nothing short of astonishing. It seems extraordinary that astronomers could have failed to spot a ninth large planet lurking inside our very own Solar System. Shocking that someone could live a relatively normal life despite missing a large chunk of their brain. Remarkable that stories first told by humans thousands of years ago could retain their accuracy to the present day. But all of these ideas are, in fact, supported by recent scientific evidence.

Each new scientific finding is amazing in its own right. Perhaps even more surprising, though, is that they all occupy a position in a single scientific landscape. At a fundamental level all of the discoveries scientists make are part of one grand, interlinked story.

Sometimes the connections between scientific discoveries are obvious: Einstein made such an enormous contribution to science that his theories provide a solid bridge that connects GPS technology to the successful efforts to measure the gravitational waves rippling through the Universe.

Sometimes the links are more obscure: the scientists working on culturing artificial meat in the lab seem to have little in common with the engineers making plans to mine asteroids in space. But in both cases the work is, in part, about securing water supplies.

According to one famous idea from the 1920s, there are no more than six social steps connecting any two people in the world. A theme explored in this book is that this "six degrees of separation" idea applies to scientific research, too. Begin with any one of the 80 stories in this book and, by following the links, it is possible to arrive at any other story in no more than six steps.

The 80 stories themselves have been drawn from across the scientific spectrum. They were chosen more or less at random from the dozens of stories that make the headlines every year. But they do share two things in common. All 80 stories describe research and developments that are no more than a few years old. And all 80 stories are about science that is, in its own way, so unexpected that it is almost unbelievable.

Colin Barras, 2018

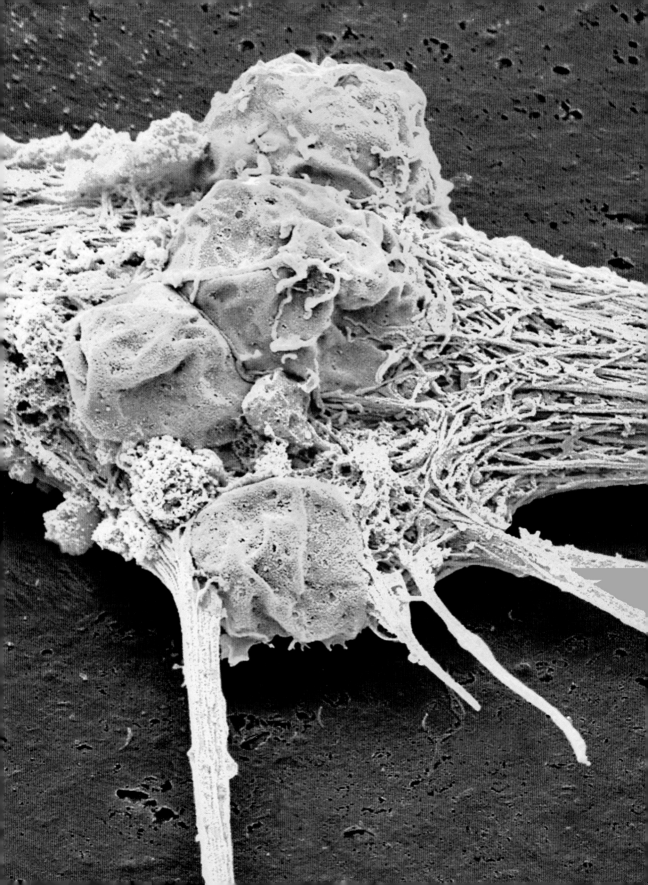

CHAPTER ONE

SPACE

THE FIRST STARS IN THE UNIVERSE

It all happened so quickly. Within a single second, space may have expanded more than a trillion trillion times; exotic subatomic particles, including the famous Higgs boson, briefly flickered into view; and temperatures dropped from hundreds of nonillions of degrees Celsius (that's a "one" followed by 30 "zeros") to a mere billion degrees.

The Big Bang took place 13.8 billion years ago.

This was the Big Bang, the birth of the Universe.

Things continued to evolve rapidly. Most of the antimatter in the known Universe disappeared. Temperatures dropped further, down to a few thousand degrees. Subatomic particles began to clump together, eventually forming the first atoms – hydrogen, helium and a tiny amount of lithium.

And then... darkness. After an initial burst of activity lasting a few hundred thousand years, the Universe entered a long and gloomy epoch – one that physicists call the Dark Ages. It would be hundreds of millions of years before the first stars shone out into space.

In 2015, astronomers found something truly unexpected: their best evidence yet that stars very like those first ones continue to shine today.

The earliest stars to shine out in the darkness of the Universe are dubbed "Population III stars." They were very different from the stars we see in the night sky today. Some theories suggest they were enormous, each with hundreds of times the mass of the Sun.

They were probably also far hotter. They burned brightly and they died young in spectacular explosions that seeded the Universe with matter to build the next generation of stars.

They died so young, in fact, that not one of these Population III stars should survive today. But perhaps, against all the odds, some of them do.

That's according to an international team of astronomers led by David Sobral, then at the University of Lisbon, Portugal. The scientists used the Very Large Telescope in the Atacama Desert of northern Chile – and several other telescopes – to survey some of the most distant galaxies from Earth.

The moon setting over the Very Large Telescope facility in the Atacama Desert of Chile.

Because the light from these distant regions of space has taken billions of years to reach us, it effectively shows us how these galaxies looked in the first billion years of the existence of the Universe.

Such distant galaxies are generally relatively dim. But some of those that Sobral and his colleagues found are unusually bright.

One in particular is three times as bright as any ancient galaxy seen before 2015. The astronomers dubbed it "Cosmos Redshift 7," or CR7 for short. The name was chosen to honor the then brightest of Real Madrid's "Galáctico" footballers, Cristiano Ronaldo, who also goes by the nickname CR7 (several of the astronomers are, like Ronaldo, Portuguese).

It's not just its brightness that makes CR7's light unusual. Its composition – or spectrum – is odd, too. Generally, the light from a galaxy betrays the presence of a vast array of elements: oxygen, iron, silicon and so on. But CR7's light contains little more than hydrogen and helium.

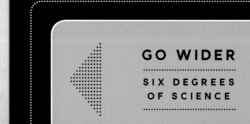

GO WIDER

SIX DEGREES
OF SCIENCE

These are the only two elements forged in abundance during the Big Bang itself, and the only two that would be found in the very earliest stars of the Universe.

If the light from CR7 contains mostly hydrogen and helium, there's a good chance that some of its stars belong in the Population III category. They might be some of the last of their kind in the known Universe.

In 2016, another team of astrophysicists studied CR7 using different telescopes – and they think they found traces of elements other than hydrogen and helium in its stars. This would suggest CR7 does not contain Population III stars after all.

CR7 was the brightest ancient galaxy astronomers had seen when it was discovered in 2015.

But the issue will probably remain unsettled for a few more years, until NASA switches on the James Webb Space Telescope. This has the power to probe deeper into space than astronomers can right now. The new telescope should be able to confirm whether or not Population III stars still lurk in those far-flung regions – and whether we really can still glimpse a type of star that burned brightly to bring the Universe's Dark Ages to an end.

For more on...
Unusual stars:

 The weirdest star in the galaxy

 How to build a star on Earth

Studying the dawn of the Universe:

 The waves that distort our planet

 Does particle physics have a problem?

 The mystery at the core of the Universe

THE END OF EVERYTHING

Out in the inky blackness of our galaxy, there are weapons of extraordinary mass destruction. And in 2013, the Internet was abuzz with rumors that one of them was pointing directly at us.

Astronomers said that sometime in the next 500,000 years – and potentially a lot sooner – our planet could be hit by a wave of radiation powerful enough to wipe out a chunk of the atmosphere. Life as we know it would end. Gamma-ray bursts are astonishingly intense explosions. The radiation from such blasts can be focused into a narrow beam that shoots out into space – and no one wants Earth to be in the firing line. The blast could destroy the ozone in the atmosphere that helps protect life from the Sun's cancer-causing ultraviolet radiation. Without that protection, many plants and animals would die.

A giant asteroid could wreak havoc if it collided with our planet.

Back in 2008, Peter Tuthill at the University of Sydney, Australia, and his colleagues were analyzing a star system called WR 104 that lies about 8,000 light years from Earth. One star in WR 104 is very old and very large. It is ready to explode, and when it does so, there is a small chance it may release a gamma-ray burst.

Worryingly, the way the star is oriented suggested to Tuthill and his fellow astronomers that Earth might be right in the firing line of the burst.

In 2013, the threat was given a greater sense of imminence. Astronomer Grant Hill at the W.M. Keck Observatory in Hawaii suggested the explosion could happen at any point in the next 500,000 years.

The threat from gamma-ray bursts should not be underestimated. In 2014, two physicists – Tsvi Piran at the Hebrew University of Jerusalem, Israel and Raul Jimenez at the University of Barcelona, Spain – concluded that gamma-ray bursts are so devastating and so common that they might well have destroyed alien life in about 90 percent of the galaxies in the known Universe.

Alien civilizations might be very rare as a consequence – which could help explain why scientific searches for extraterrestrial intelligence have so far drawn a blank.

Even in the few galaxies that could harbor life – and our own Milky Way galaxy is obviously one of them – it's only in the outer regions that gamma-ray bursts are rare enough to allow life to survive. According to Piran and Jimenez, if Earth was a little closer to the Milky Way's heart, there's a good chance it would be lifeless.

Unfortunately, this doesn't mean Earth is immune to the threat of gamma-ray bursts – just that about 500 million years might pass without the planet being struck by one. Given that our planet is about 4.54 billion years old, it must have been hit by several in the past.

In 2004, Adrian Melott at the University of Kansas in Lawrence and his colleagues suggested that the last gamma-ray burst struck Earth about 440 million years ago – and triggered one of the biggest mass extinctions of life that our planet has ever experienced. Little wonder that 2013's news about WR 104 worried many.

Fortunately, subsequent research has brought better news. The latest assessment suggests the dangerous star in WR 104 is not "pointing" directly at us after all. So, even if it releases a gamma-ray burst in the future, it should miss Earth.

Many marine species vanished at the end of the Ordovician period – perhaps they were wiped out by a gamma-ray burst?

We probably shouldn't breathe too easily, though. Gamma-ray bursts are just one of the alien threats to life on Earth.

Asteroids pose their own risk. A large space rock almost certainly contributed to another of the biggest mass extinctions – the one that killed off the large dinosaurs about 66 million years ago. Even rogue stars wandering through the cosmos could prove deadly. This threat was brought into focus in 2015 when scientists discovered that one such star passed through the outer edges of the Solar System just 70,000 years ago, around the time that our species was beginning to leave Africa and spread across the world.

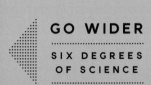

GO WIDER

SIX DEGREES OF SCIENCE

For more on…
Extraterrestrial threats to life:

How to get rich in space

The dwarf dinosaurs of Transylvania

Earth-based threats to life:

The magma that could kill us all

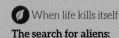
When life kills itself

The search for aliens:

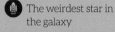
The weirdest star in the galaxy

THE EXTRAORDINARY STORY OF PLANET NINE

It has ten times the mass of our planet, twice its diameter – and, as of early 2018, it's never been seen. Somewhere in the furthest reaches of our Solar System, a brand new planet may lurk.

Is there a ninth large planet in the furthest reaches of the Solar System?

If so, artists' impressions of the Solar System will need to be updated.

Astronomers seem to find new planets on an extraordinarily regular basis. Thousands have been spotted since the late 1980s. But these are all "exoplanets" – worlds orbiting distant stars.

Planet Nine is different. This proposed planet orbits our very own Sun, placing it virtually on our doorstep. It would become only the third new planet discovered in our Solar System since ancient times – the first discovery of its kind in about 170 years. It's close enough that some physicists think humans could begin exploring the planet within a few decades, using unmanned spacecraft.

That's assuming it really exists.

In January 2016, two astronomers – Konstantin Batygin and Michael Brown at the California Institute of Technology in Pasadena – said Planet Nine is real. Their evidence came from the unusually coordinated alignment of icy debris in the Kuiper Belt, a region of space junk that lies on the outer edges of the Solar System, far beyond the eight known planets.

Computer simulations showed that the best way to explain the alignment of this debris would be to plonk a hypothetical large planet somewhere in the vicinity – one with a gravitational pull that would interfere with all of the icy debris in a similar way.

Neat though this scientific reasoning is, it's not enough on its own to convince the scientific world that Planet Nine exists. However, Batygin and Brown's computer simulations did more than just account for the observed strange alignment of the icy debris. They predicted that Planet Nine would interfere in a very characteristic way with the orbits of other chunks of debris in the outer Solar System.

When the two scientists looked at astronomical data, they discovered that these predictions were correct – all

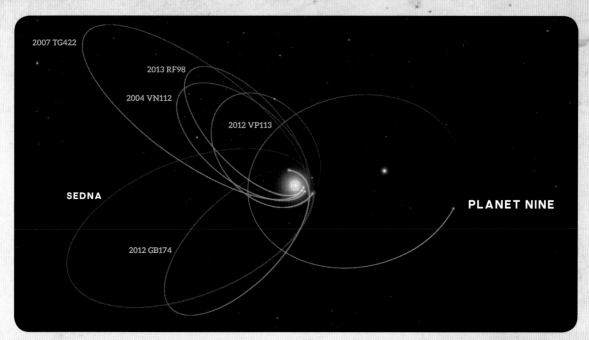

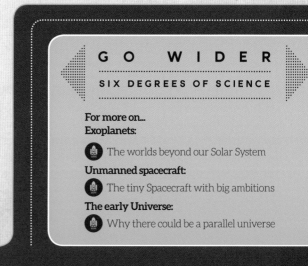

2007 TG422

2013 RF98

2004 VN112

2012 VP113

SEDNA

PLANET NINE

2012 GB174

sorts of objects in the outer Solar System are behaving just as they should do if there is a Planet Nine. It was this discovery that helped persuade other (but not all) astronomers that Planet Nine must be real.

After the January 2016 announcement, planetary scientists began busily working on theories to explain Planet Nine's history. One idea was that the hypothetical world is actually a planet "stolen" from another star early in the life of the Solar System – which would technically make Planet Nine an exoplanet after all.

By October 2017, NASA said there was so much evidence for the existence of Planet Nine that it is difficult to imagine the Solar System without it. The mysterious planet may be a "super Earth." These are rocky planets that are much larger than Earth, and astronomical surveys suggest they are common in other star systems.

But despite the intense interest, the new world hadn't turned up on any telescope images as of early 2018. Why is Planet Nine proving so elusive?

Part of the problem is that Planet Nine is calculated to be so far from the Sun that it receives almost no sunlight. This means it probably doesn't reflect much back into space – so, unlike many of the other planets, it won't show up as a nice bright point in the night sky.

But it might be emitting signals that scientists can detect. Some astronomers think that radio waves – which,

Some of the most distant known objects in the Solar System have strangely aligned orbits (marked in purple): the orbit of a large ninth planet could explain why.

confusingly, are essentially a form of light – might help locate Planet Nine. A class of telescopes that detects radio waves from the faint afterglow of the Big Bang could be recruited to look for telltale signs of Planet Nine's existence.

Such is the unpredictable nature of science: in 2017, telescopes designed to study the earliest history of the known Universe were being enlisted to help locate the latest member of our planetary family.

GO WIDER

SIX DEGREES OF SCIENCE

For more on...
Exoplanets:
The worlds beyond our Solar System

Unmanned spacecraft:
The tiny Spacecraft with big ambitions

The early Universe:
Why there could be a parallel universe

THE TINY SPACECRAFT WITH BIG AMBITIONS

For a billionaire, virtually anything is possible – even literally reaching for the stars. Yuri Milner, a venture capitalist and physicist, announced in 2016 that he wants to send a fleet of spacecraft some 40 trillion kilometers (25 trillion miles), or 4.3 light years, through space to a nearby star system.

It's part of an ambitious program – "Breakthrough Initiatives" – that Milner is funding to search for extraterrestrial intelligence. The spacecraft plan, dubbed Breakthrough Starshot, has been gaining all the attention so far. It's humanity's first real attempt to stride out into the interstellar space beyond our Solar System.

But this first giant leap into the cosmos will leave a very small footprint. Milner's proposed "StarChips" – which researchers hope could be ready to launch in a few decades – will each be small enough to sit in the palm of your hand.

Size and mass are the enemy when it comes to accelerating at super-fast speeds. Using current technology, standard spacecraft are just too big and heavy to accelerate to the speeds necessary for interstellar travel. Spacecraft the size of a smartphone, though, are small enough.

To make the job of accelerating such tiny spacecraft even easier, the plan is to equip each StarChip with a "light sail" – a gossamer-thin sheet built from a lightweight polymer that can be unfurled in space. Zapping a light sail with a high-power laser array on Earth would give a StarChip enough of a kick to accelerate to one-fifth the speed of light in a matter of minutes. Then the StarChip could simply coast through the emptiness of space.

Famous physicist Stephen Hawking, a Breakthrough Initiatives board member, calls the plan a very exciting first step on our journey to the stars. It's a journey he thinks is vital if human civilization is to survive in the long term, given that Earth might one day be hit by a large space rock, or a deadly burst of radiation powerful enough to wipe out most life as we know it.

(Top) Solar sails – like this one, carried by a space probe launched in 2010 – could send spacecraft into interstellar space.

(Above) Yuri Milner and Stephen Hawking outline the Breakthrough Starshot plans.

The year 2016 brought exciting news for those, like Hawking, who think humans will soon need to colonize alien worlds.

Astronomers have now identified thousands of planets orbiting other stars in our galaxy, but all of these "exoplanets" are so far away from Earth that humans are

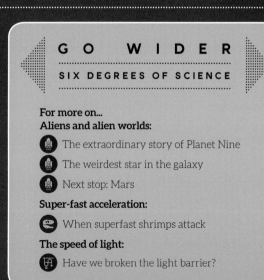
Could a swarm of tiny "StarChips" find signs of life on Proxima b?

unlikely to develop the technology to reach them with probes in the foreseeable future.

All, that is, except for Proxima b.

The European Southern Observatory announced the discovery of this exoplanet in August 2016. Proxima b orbits Proxima Centauri, which is one of the three stars in the Alpha Centauri star system – the closest star system to our Solar System. Initial investigations suggest Proxima b might have the right conditions to be habitable. By October 2016, there were reports that the planet might even be home to Earth-like oceans.

It just so happens that the Alpha Centauri star system is exactly where Milner and his team of scientists were already planning to send their StarChips.

Light – which moves faster than anything else in the Universe – can make the journey from Earth to Alpha Centauri in just over four years. The proposed Breakthrough Starshot spacecraft could cover the distance in 20 or 30 years. That's an astonishingly rapid journey when you consider that sending humans to Mars – about 500,000 times nearer to Earth than Alpha Centauri – is expected to take eight months on current technology.

In June 2017, Breakthrough Starshot launched several prototype StarChips into orbit around Earth. In another 50 years, real StarChips might be beaming back images

of Proxima b and its hypothetical oceans. When those images arrive, they may even provide the first irrefutable evidence of alien life.

That's assuming we don't find convincing evidence of aliens on the surface of Mars, or around other distant stars, before that time.

GO WIDER

SIX DEGREES OF SCIENCE

For more on...
Aliens and alien worlds:

The extraordinary story of Planet Nine

The weirdest star in the galaxy

Next stop: Mars

Super-fast acceleration:

When superfast shrimps attack

The speed of light:

Have we broken the light barrier?

HOW TO GET RICH IN SPACE

It's been described as a game changer for space exploration: a bold vision to chase down asteroids flying close to Earth and strip them of their natural resources. And what will be the most valuable commodity on offer in this new realm of space mining? Not gold, silver or diamonds.

The ultimate prize is something that Earth is already literally awash with: water.

In recent decades, asteroids have been viewed as a threat to life on Earth. The hazard appeared all the more serious when evidence emerged in the 1990s that an asteroid at least 10 kilometers (6 miles) in diameter slammed into the Gulf of Mexico about 66 million years ago. The "Chicxulub asteroid" almost certainly played a big part in a mass extinction that claimed the lives of all the large dinosaurs. Large lumps of space debris could easily claim lives today if they hit our planet.

However, more recently attitudes have begun to shift.

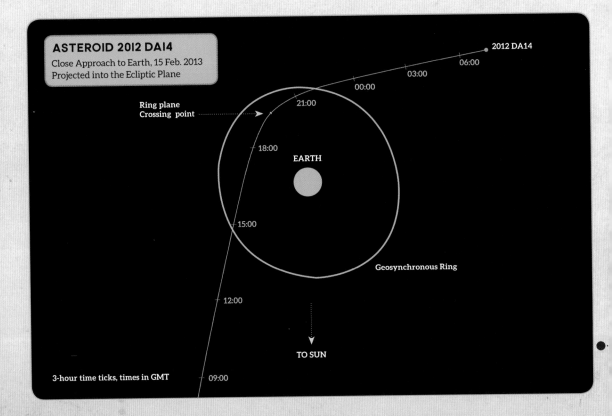

ASTEROID 2012 DA14
Close Approach to Earth, 15 Feb. 2013
Projected into the Ecliptic Plane

Ring plane
Crossing point

2012 DA14

06:00

03:00

00:00

21:00

18:00

EARTH

15:00

Geosynchronous Ring

12:00

TO SUN

09:00

3-hour time ticks, times in GMT

In February 2013, for instance, a 30-meter- (90-foot-) wide asteroid called 2012 DA14 passed perilously close to Earth – nearer, in fact, than some of the communications satellites that orbit our planet. Before it was clear that the rock would miss Earth, some observers worried about the threat it might pose.

The mining industry took a different view. Always on the lookout for ways to exploit modern technology for mining activities, they saw opportunity, not crisis. One firm – Deep Space Industries – claimed 2012 DA14 carried natural resources worth an estimated $195 billion (about £160 billion). The space miners, still years away from actually being able to exploit asteroids, could only watch helplessly as the valuable rock whizzed by our planet and vanished once more into the depths of space.

The resources carried in 2012 DA14 included metals that are vital for the technology industry, but difficult for manufacturers to source here on Earth. Market supplies of the indium in smartphone touchscreens, for instance, or the neodymium in smartphone earbuds – and in wind turbines – might begin to dry up in the next few years, according to some assessments.

However, 2012 DA14's estimated price tag also included $65-billion- (£52 billion-) worth of water.

There are a few simple reasons why a commodity that is so abundant on Earth fetches such a high price in space. Most importantly, water is very expensive to transport off the planet, which instantly makes natural sources in space itself much more valuable.

A near miss: in 2013 one asteroid passed inside Earth's ring of geosynchronous communications satellites.

(Above) The Rosetta mission proved that it is now possible to land on a comet.

Then there is the fact that water is incredibly useful for the space industry. Split it into its constituent elements and water can become a powerful rocket fuel. In its natural state, it makes a great radiation shield that could protect astronauts in space and in bases on Mars or other planets. By early 2017 there were media reports that the oil states of the Middle East recognize the potential: they have begun investing in companies that could one day be mining for water in space.

A few years ago, it might have seemed unlikely that space mining could work in practice. But in 2014, the European Space Agency's Rosetta mission managed to land a module on the 4-kilometer- (2.5-mile-) wide 67P/Churyumov–Gerasimenko comet, demonstrating that it is now possible to rendezvous with – and land on – small asteroids and comets traveling through the Solar System.

In 2015, engineers announced a realistic proposal to mine water from an asteroid. They think they can use a technique called optical mining to drive 100 tonnes (110 tons) of water out of an asteroid and into a large inflatable bag that could then be towed to some sort of space base. Best of all, the approach would require a single launch of the commercial SpaceX Falcon 9 rocket, making it a relatively cheap way to mine water in space.

Human space exploration might hang on the success of such ventures. In the arid wasteland of space, little is more valuable than water.

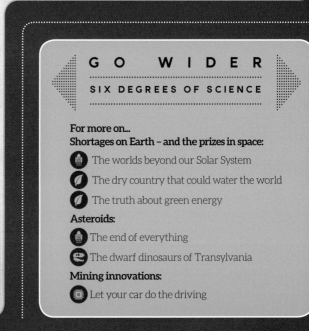

GO WIDER

SIX DEGREES OF SCIENCE

For more on...
Shortages on Earth – and the prizes in space:

The worlds beyond our Solar System

The dry country that could water the world

The truth about green energy

Asteroids:

The end of everything

The dwarf dinosaurs of Transylvania

Mining innovations:

Let your car do the driving

WHY THERE COULD BE A PARALLEL UNIVERSE

You might not be the richest person on Earth – or the most famous, or the most intelligent. But according to some physicists, there is another Earth out there in which you are all three. Welcome to the Multiverse – one of the most mind-bending realms of modern science.

The idea is simple enough: some scientists are convinced that the known Universe is just one of many parallel universes out there. Together, all of those parallel universes form the Multiverse.

Appropriately enough, there is not one Multiverse theory but several. One of these theories assumes that the Universe must contain advanced alien civilizations with the computing technology to "re-run" the Universe in super-realistic virtual reality. Each of those simulations is, in effect, a parallel Universe.

Slightly unsettlingly, the scientists behind this particular theory say there is a good chance we are actually living in one of these simulations. They point out that there should be countless simulated universes running on alien computers – and only one "real" Universe. Given the odds, it's far more likely we are in one of the simulated versions.

Not all Multiverse theories rely on virtual reality and aliens, though. One theory begins with scientists' efforts to study the earliest moments of the known Universe.

Most scientists accept the idea that our Universe began with the Big Bang. Although this event occurred 13.8 billion years ago, its faint afterglow survives to this day – it's called the Cosmic Microwave Background (CMB).

A class of telescopes that detect radio waves has been looking at this CMB for a few decades, and it's revealed something very odd. The CMB on one side of the known Universe is almost exactly the same temperature as the CMB on the opposite side.

That shouldn't be the case.

A diagram of the evolution of the universe, from the Big Bang (top) to now. Looking back on our galactic history may help scientists discover possible parallel universes.

Given enough time, heat can gradually spread through a volume to give it the same temperature throughout. But these two regions of space are so far apart that nothing – not even light – has had time to travel between them since the Universe began. They simply shouldn't be at the same temperature. But they are. Why?

The best way to explain this is to introduce a concept that scientists call cosmic inflation. According to some physicists, within the first second of the Big Bang, space expanded at a quite extraordinary rate – faster even than the speed of light. The process means everywhere in the known Universe had the opportunity to communicate right at the start of everything, which helps explain why the CMB looks the same everywhere scientists look.

However, the curious thing about cosmic inflation is that physicists have no way of knowing where – or even if – it ended. In principle it didn't, which means that space might be infinite.

The matter that forms galaxies, planets, and – potentially – people, will be arranged in different configurations elsewhere in the infinity of space. But fundamentally, there is a finite number of atoms in our observable Universe – roughly one thousand quadrillion vigintillion (a "one" followed by 80 "zeros") according to some estimates. There are only so many ways all of those atoms can be arranged.

If you travel far enough through space, simple probability says you will eventually come across a region where the atoms happen to be arranged in an almost exactly identical way to the configuration of our known Universe. There will be a twin Sun, a twin Earth, and a twin you.

And it doesn't stop there. In theory, there could be an infinite number of these twin Earths out there. History might have played out slightly differently on each one – hence the possibility that we are each kings of our own Earth somewhere in the Multiverse.

All of this may sound more like science fiction than science fact, and many scientists agree. They are uneasy

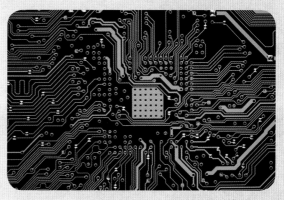

(Above) How far through space would we have to travel to find Earth's twin?

(Below) Powerful alien supercomputers could be running virtual simulations of the Universe.

about pushing the idea of the Multiverse too far precisely because it's very difficult to test the idea that these parallel Earths really exist.

But, if there really are a multitude of Earths out there, perhaps there's one where scientists have somehow found a way to prove the existence of the Multiverse. It just hasn't happened on this Earth. Yet.

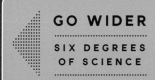

GO WIDER

SIX DEGREES
OF SCIENCE

For more on...
Advanced alien civilizations:

 The weirdest star in the galaxy

Studying the early Universe:

 The extraordinary story of Planet Nine

The waves that distort our planet

THE WEIRDEST STAR IN THE GALAXY

Deep in our galaxy, there's a star that's behaving erratically. While most other stars shine steadily, this one – KIC 8462852 – is brightening and dimming in a very peculiar way.

Its behavior is so peculiar, in fact, that in 2015 some scientists began to entertain an extraordinary idea. KIC 8462852, they said, might be betraying the presence of an alien civilization vastly more advanced than ours.

KIC 8462852 – also known as "Tabby's Star" – is about 1,280 light years from Earth. This means it's too far away to see with the naked eye. But it has been turning up on telescope images since the 1890s. For most of that time, it attracted little scientific attention.

That changed in 2015. Astronomers and citizen scientists poring through data captured by Kepler, a space telescope that searches for alien planets, noticed that the light coming from KIC 8462852 changed in an irregular pattern between 2009 and 2013.

The scientists began searching for an explanation. Their initial suggestion, in 2015, was that there must be a dense cloud of comets orbiting the star and blocking some of the light. Such a comet cloud might be irregular in size and shape, so as it passes across the face of the star it would interfere with the starlight in an unusual way, leading to the odd brightness pattern.

However, it soon became clear that this explanation didn't quite fit. KIC 8462852 is a relatively old star – and it's very unusual for mature stars to be surrounded by dense comet clouds.

That left room for an alternative. Later in 2015, a team of astronomers led by Jason Wright at Pennsylvania State University in University Park suggested one. They considered the possibility that the star might in fact be surrounded by purposely built "space megastructures" that blocked its light in an unusual way. The handiwork of a super-advanced alien civilization, Wright and his

(Above) Advanced alien civilizations might have the capacity to build structures around stars and harness solar energy.

Swarms of comet fragments offer a more mundane explanation for the behavior of KIC 8462852.

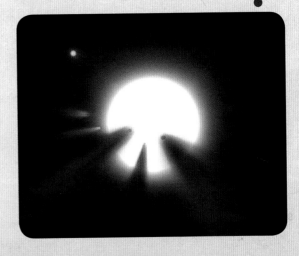

KIC 8462852

colleagues suggested, might explain KIC 8462852's extraordinary behavior.

In fairness, the "alien civilization" idea has always been seen as an explanation of last resort – an option to turn to when all other explanations have been found wanting. It's unlikely an alien civilization could have grown advanced enough to build large structures around a star – not least because space hazards like gamma-ray bursts pose a deadly threat to life in many parts of our galaxy.

But as more was learned about KIC 8462852, the star only looked stranger – which made the alien civilization idea begin to look surprisingly plausible.

By January 2016, a careful search through astronomical archives suggested that KIC 8462852 had actually behaved erratically ever since it was first discovered in the nineteenth century. It had dimmed gradually, by about 15 percent, over that time.

Only in September 2016, did a natural, non-alien explanation begin to look more likely again. Some astronomers found hints of another star directly in line between Earth and KIC 8462852 – and this star also seems to be behaving strangely. This suggests that there might be fields of space junk in the vastness of interstellar space between both stars and our Earth. Star KIC 8462852 itself might not be quite as strange as it first appeared.

A NASA study published in October 2017 concluded as much too: ultraviolet light from the star dims more than infrared light, which is what we would expect if the dimming was due to dust rather than an alien megastructure.

In other words, it might be back to Square One in the search for alien life, let alone alien civilizations. Short of finding messages that aliens encoded into bacterial

The oddly behaving star KIC 8462852, imaged in October 2015.

DNA long ago – probably a long shot – our best chance of detecting extraterrestrials might come from studying potentially habitable planets around stars far closer to Earth than KIC 8462852.

Proxima b, found in 2016, is hundreds of times nearer to our planet, making it easier to study. It could be reached in a few decades with a revolutionary new class of tiny spacecraft – and it could harbor life.

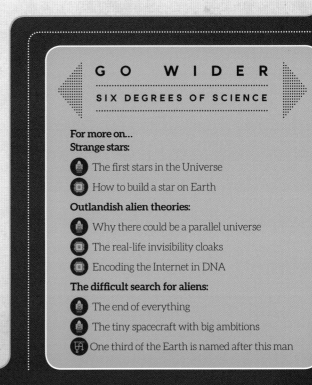

G O   W I D E R

SIX DEGREES OF SCIENCE

For more on...
Strange stars:

The first stars in the Universe

How to build a star on Earth

Outlandish alien theories:

Why there could be a parallel universe

The real-life invisibility cloaks

Encoding the Internet in DNA

The difficult search for aliens:

The end of everything

The tiny spacecraft with big ambitions

One third of the Earth is named after this man

NEXT STOP: MARS

They already walk among us. In 2014, NASA suggested that the first people who will visit Mars have now been born. The trip they will take one day will be extraordinary: exhilarating, a one-of-a-kind experience, and a golden opportunity to discover alien life. It will also be incredibly boring.

NASA's current plan is to have humans touch down on Mars sometime in the 2030s – although in July 2017 the space agency admitted that the costs involved might push back that date.

The groundwork for that first manned mission has already begun. Robotic rovers now on the Red Planet are monitoring radiation levels so that the space agency can develop equipment tough enough to keep humans alive and well on the alien world. Astronauts on the International Space Station are testing the communications systems for deep space missions. And biologists on Earth are monitoring those same astronauts to better understand how human health is affected by a prolonged spell in space – including whether the microbes that live in the human gut flourish or perish.

On current technology, the trip to the Red Planet will take about eight months each way. NASA is fully aware that such a long journey will impose a significant psychological burden. So much so, in fact, that in 2014 the space agency funded a preliminary study to investigate the idea of putting humans into a state of deep sleep for the journey.

Perhaps the idea was to learn from animals like the Arctic ground squirrel that allow their bodies to cool – or even freeze – before entering months of hibernation. But it will be years before the biological lessons learned from studying these specialized animals can be applied to humans. NASA didn't pursue the "deep sleep" idea.

Instead, in 2015, it announced hopes to develop new propulsion technology that could cut the journey time to Mars in half. A shorter trip would reduce the amount of

Is the Martian surface actually hospitable enough to welcome human visitors?

Another day on "Mars" – actually the rocky surface of a Hawaiian volcano.

precious water and other resources the mission would have to carry.

However, even if the space agency succeeds in developing a spacecraft that can reach Mars in four months, that's still a very long time for a crew to live in a relatively small space with no possible means of escape.

How will they cope? Psychologists are trying to find out. There have already been several successful "Missions to Mars" – all carried out right here on Earth.

In June 2010, six men began a 520-day-long simulated Mars mission in a facility in Moscow. They spent months on end in a mock spacecraft before "landing" and exploring a simulated Martian landscape. Then, exploration complete, it was back inside the mock spacecraft for another eight months of "travel" back to Earth.

Throughout, the six men completed questionnaires to gauge their psychological state and assess whether the cramped conditions would lead to stress and arguments. The results suggested a crew could cope with the lack of privacy and the cramped and boring environment.

Another year-long "mission," this time in a solar-powered dome in Hawaii, ended in August 2016. Again, it was successful – although those involved confirmed that boredom was one of their biggest enemies.

Some critics argue that simulations like these are too safe. The "crews" know full well that they can abort at any time – Earth is always just a closed door away. Would their psychological response be different if they knew that option didn't exist? And if the "mission" itself carried a real risk of death?

Simulated Mars missions in extreme environments might help fill these knowledge gaps. NASA is now using bases in Antarctica and even deep below the Atlantic Ocean to study how humans cope when forced to live for days on end in cramped and confined conditions in a truly hostile environment.

The lessons from those experiments are still being learned. But whatever the findings, it's a fair bet that there will be psychologists on hand at mission control when the first manned mission to Mars blasts off.

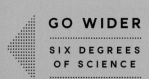

GO WIDER

SIX DEGREES OF SCIENCE

For more on...
New space technology:

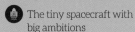 The tiny spacecraft with big ambitions

The life-saving power of excrement

Biology and medicine with space applications:

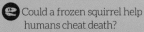 Could a frozen squirrel help humans cheat death?

The dangers of boredom:

 Can boredom be fatal?

THE WAVES THAT DISTORT OUR PLANET

On Monday, September 14, 2015, just before 11 a.m. UK time, our planet very briefly distorted. A subtle wave rippling through space itself washed over and through Earth, very slightly stretching and then compressing the atmosphere, the oceans, the continents – and living things.

Earth had just experienced a gravitational wave. And for the first time in history, scientists detected the event.

Ever since Einstein, scientists have come to think of time and space as combining to form a single entity that they call "space-time." One of the predictions that stem from Einstein's theories of relativity is that a sudden jolt somewhere in the Universe would send "gravitational waves" rippling through this space-time. Some physicists compare these waves to the ripples on a millpond after someone has thrown in a stone.

In the vastness of space, it's when two black holes collide that some of the largest gravitational waves should be generated. It was the waves from one such collision – between two black holes 1.4 billion light years from Earth – that were detected in September 2015.

As the gravitational waves radiated away from the colliding black holes, they stretched and compressed space-time: the distance between two fixed points expanded and then contracted as each wave washed over them.

The trouble is that the further the waves traveled, the weaker they became – again just like ripples on a pond. By the time the ripples from the distant black hole collision washed over Earth, they were barely stretching and squeezing space-time at all. They warped space-time not by meters, centimeters or even millimeters, but by distances on a scale far smaller than the size of individual atoms.

This is why it took scientists decades to detect gravitational waves. They needed to build a system of lasers sensitive enough to detect the tiny distortions in space-time that occur when these very weak waves wash across the Earth.

The detectors that finally proved up to the job are called the Laser Interferometer Gravitational-wave Observatory

When two black holes spin into one another, the collision sends waves radiating out through the fabric of space.

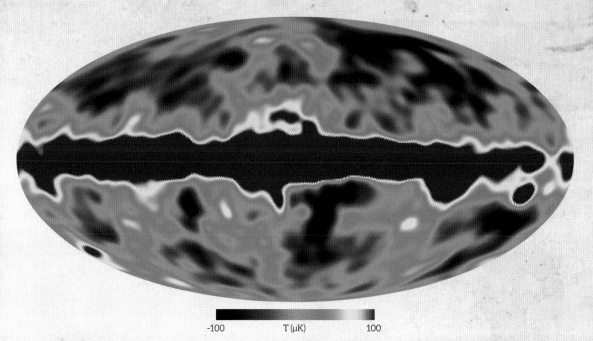

-100   T (μK)   100

The cosmic microwave background of the universe holds clues for understanding its birth at the Big Bang. T=temperature in microkelvin.

(LIGO) instruments. There are two of them – one in the US state of Louisiana and one 3,000 kilometers (1,865 miles) to the northwest in Washington state. Because gravitational waves are theorized to travel at the speed of light, it should take about 10 milliseconds for a wave to travel between the two observatories. If both laser systems detect a similar disturbance, one 10 milliseconds before the other, it's good evidence that a gravitational wave has just swept through Earth. This is what happened on September 14, 2015.

More gravitational waves have been detected since 2015. In August 2017, for instance, astronomers detected gravitational waves and a powerful gamma ray burst from a "kilonova" – an event caused by the merger of two neutron stars.

Such successes are a big deal. Not only do gravitational waves provide yet more confirmation of Einstein's ideas about the nature of space and time, they offer scientists a brand new way to study the future of the Universe – and its past.

Gravitational waves should give scientists a new tool for measuring how fast the Universe is expanding. As importantly, the waves might provide a glimpse further into the Universe's past than scientists have ever peered before.

Some scientists think the Universe expanded at a phenomenal rate within the first second of the Big Bang – an idea called cosmic inflation. Evidence of this inflation is found in the uniform appearance of the cosmic microwave background (CMB), the faint afterglow left by the Big Bang.

Cosmic inflation should also have generated powerful gravitational waves and – extraordinary as it may seem – the remnants of those primordial gravitational waves should still be out there in the CMB, just waiting to be measured.

GO WIDER

SIX DEGREES OF SCIENCE

For more on...
Einstein's bold ideas:

 Why Earth's core is younger than you think

Have we broken the light barrier?

The expansion of the early Universe:

The first stars in the Universe

Why there could be a parallel universe

THE WORLDS BEYOND OUR SOLAR SYSTEM

The sunsets must be spectacular on Kepler-1647b. NASA scientists announced the discovery of the Jupiter-like planet in June 2016. It's about 3,700 light years from Earth and it takes 1,107 days – rather than Earth's 365 days – to travel around its sun. Or, rather, its suns. Kepler-1647b orbits two stars, not one. At dusk, there are two suns setting in its sky.

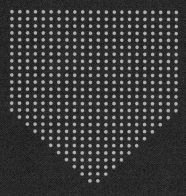

Kepler 16b was the first planet confirmed to be in orbit around two stars.

Kepler-1647b is one of a mere handful of "circumbinary" planets (planets that orbit two stars) that have been discovered since 2011. The worlds are sometimes nicknamed "Tatooine" planets, after the fictitious circumbinary world from the *Star Wars* movies. The discovery of such exotic worlds shows that, in the vastness of the known Universe, truth can be just as strange as fiction.

Take 55 Cancri e, also known as Janssen. It was discovered in 2004, but made the news in 2012 after a reassessment suggested it could be composed largely of carbon. Because of the temperatures and pressures inside the planet, as much as one-third of this world could be in the form of diamond. Its monetary value? Perhaps as much as $26.9 nonillion – about £22 nonillion (one nonillion is a million trillion trillion).

It's a reminder of the valuable resources on offer in space. That said, given that the diamond planet lies 40 light years from Earth, even the most ambitious space-mining projects won't be exploiting it any time soon.

Then there's TrES-2b, a planet discovered in 2011 that is darker than dark. It's another Jupiter-like planet – but unlike Jupiter, which reflects light from the Sun to shine brightly in our night sky, TrES-2b reflects only one percent of the light from its star. It's a black sphere. Planetary scientists still have no idea exactly why.

Gliese 1214 b, meanwhile, is a planet a little more like Earth in size – but potentially radically different from our home world in appearance. Discovered in 2009, this planet has an estimated density so low that it might be composed mostly of water. Scientists analyzed its atmosphere in 2010: it could contain little more than steam, consistent with the idea that Gliese 1214 b is a real-life waterworld.

PSR B1620-26 b is worth a mention, too. This world has been nicknamed the "Genesis Planet," because of its astonishing age. Astronomers think it might be about 12.7 billion years old, making it one of the oldest exoplanets found to date. It is about 12,400 light years from Earth – and, like Kepler-1647b, it orbits two stars.

Sadly, all of these exotic exoplanets are much too far away for humans to explore with space probes in the

Some alien worlds may be built from diamond.

(Opposite) The ancient "Genesis Planet," as it might appear from the twisted surface of its hypothetical moon.

foreseeable future. But two are not. Early in 2016, planetary scientists announced evidence that our Solar System might be home to a mysterious ninth planet – Planet Nine. As of late 2016 it had not actually been seen – but some scientists suggest the strange planet may be an exoplanet that our Sun somehow stole from another star.

Later in 2016, meanwhile, another team of planetary scientists announced the discovery of Proxima b, an Earth-like exoplanet orbiting a star just 4.2 light years from Earth. Reports in October 2016 suggested it might have oceans, making it suitable for alien life.

Whether or not aliens do live on exoplanets like Proxima b might prove difficult to demonstrate. Astronomers "sniff" the atmospheres of such planets for ozone, a byproduct of Earth-like life. But an October 2017 study suggested that because Proxima b orbits its star so frequently – about every 11 Earth days – airflow in its atmosphere might concentrate the ozone at the equator where it is more difficult for us to detect.

GO WIDER

SIX DEGREES OF SCIENCE

For more on...

Exploiting resources in space:

 How to get rich in space

The exoplanet on our doorstep:

The extraordinary story of Planet Nine

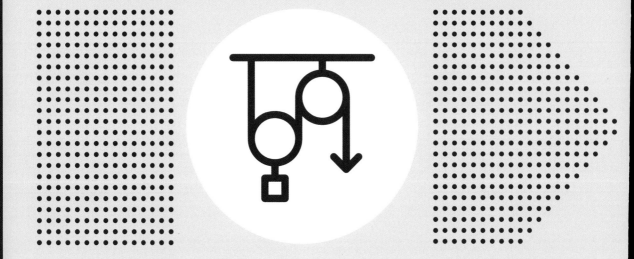

CHAPTER TWO

PHYSICS

WHY EARTH'S CORE IS YOUNGER THAN YOU THINK

Our Earth is a mind-boggling 4.54 billion years old. At least, it is at the surface. The rocks and minerals at the center of the Earth, on the other hand, are very slightly more youthful. They formed at the same time as the rest of the planet, but recent calculations suggest Earth's core is nevertheless about two-and-a-half years younger than the ground immediately beneath your feet.

This weird paradox is all tied up in Einstein's theories of relativity – ideas which play a vital role in the smooth running of consumer GPS systems.

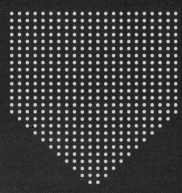

Earth's core is aging more slowly than the planet's surface.

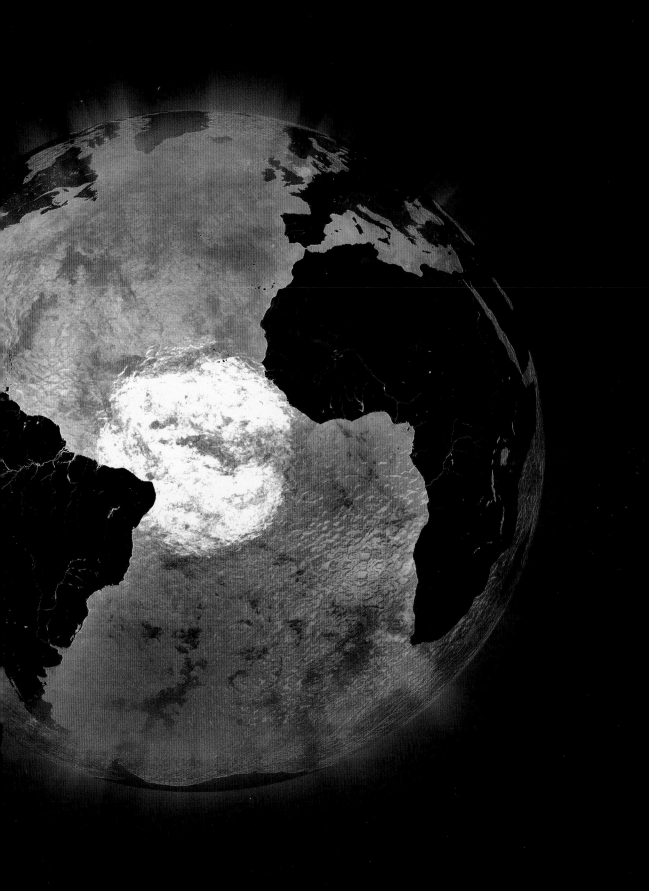

There's no easy way to explain relativity. It takes years to fully understand Einstein's ideas. But they carry a simple message that is easy to understand: reality is a lot stranger than it might appear.

Our everyday lives tell us that certain things are reassuringly constant. An hour always takes exactly the same amount of time to pass, for instance, and the distance between two stationary objects never changes. In reality, both space and time are rendered malleable by the need to maintain a constant speed of light everywhere in the known Universe. Strange as it seems, space and time are readily stretched and squeezed, depending on your position in the Universe and the speed at which you are moving.

Over the course of the average human lifespan on Earth, these differences don't add up in a way we would notice, although the differences are there nonetheless. For instance, when astronaut Scott Kelly began a year-long mission aboard the International Space Station (ISS) in 2015, he was six minutes younger than his identical twin brother, Mark (also a former astronaut). At the end of the mission, the age difference between the twins had changed by a few milliseconds.

Over the course of the Earth's 4.54 billion year lifespan, the differences add up in a more dramatic way.

Since the second half of the twentieth century, physicists have known that time has run at a slower rate in the Earth's core relative to the crust, due in part to differences in the pull of gravity at depth. Richard Feynman, a great American physicist, discussed the phenomenon in a series of lectures in the 1960s. He said that Earth's core should be a day or two younger than the surface by now.

No one thought to check Feynman's math until 2016. Then, physicist Ulrik Uggerhøj at Aarhus University in Denmark and his colleagues ran the calculations for themselves – and found that Feynman's oft-quoted figure is wrong. The core isn't two days younger than the crust; it's more than two years younger.

It's easy to dismiss this sort of science as abstract – after all, none of us will ever see the deep interior of our planet. But relativity is anything but arcane: adjusting for the effects of relativity ensures the GPS navigation apps built into smart phones and tablet PCs continue to work as they should.

The GPS system is based on a series of very accurate clocks in satellites orbiting Earth. Those clocks tick at a different rate in orbit than they would at ground level – just as time passes at a slightly different rate for astronauts on the ISS. That's clearly a problem: GPS receivers in smart phones and other devices rely on the clocks in orbit retaining their accuracy.

This is where Einstein's theories of relativity can help. They tell physicists that time passes slightly quicker on the satellites, because they are moving at high speed. And, confusingly, they also tell physicists that time passes slightly more slowly on the satellite because the pull of gravity is different in orbit.

Taking both sources of error into account, scientists can calculate exactly how much the GPS clocks will drift each day, and compensate for the error. Doing so keeps the GPS system working accurately day after day, year after year.

On the face of it, Einstein's theories of relativity might seem to have little impact on everyday life. But without them, many of us would be lost.

GO WIDER

SIX DEGREES
OF SCIENCE

For more on...

Einstein's predictions:

 The waves that distort
our planet

Earth's mysterious interior:

One third of the Earth is
named after this man

Versatile smart phones:

The end of the audio jack

DOES PARTICLE PHYSICS HAVE A PROBLEM?

It was a discovery that looked set to revolutionize what physicists know about the subatomic world. In 2015, two of the large experiments at the famous Large Hadron Collider (LHC) that straddles the Swiss-French border both reported tantalizing signs of a mysterious new subatomic particle, the existence of which no one had predicted.

It took ten years to build the Large Hadron Collider on the Swiss-French border.

Physicists went wild, dashing out hundreds of new scientific papers with detailed theories to try to explain the identity and significance of the new particle. Early in 2016, excitement was sky high.

Then sobering reality crashed the party. By August 2016, the two LHC experiments had analyzed more data – and evidence of the enigmatic particle had vanished. The earlier signal had been a statistical anomaly, the equivalent of a run of several "heads" when tossing a coin.

So what do the physicists who study these subatomic particles do now?

Many people will be familiar with the idea that our world is built from atoms. But atoms themselves are built

from smaller "subatomic" particles – electrons, protons and neutrons. Protons and neutrons are, in turn, built out of another class of even smaller subatomic particles named "quarks."

In fact, there is a huge variety of subatomic particles, and physicists are still trying to identify and understand them all. They build very large and very expensive particle colliders to do so. Inside these colliders, subatomic particles like protons are smashed into each other at extremely high speeds, concentrating energy in a tiny space and

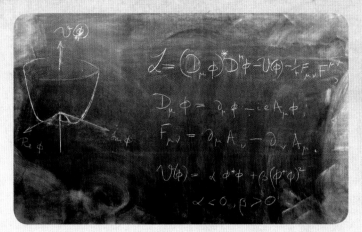

briefly recreating conditions last seen at the dawn of the Universe. Those conditions allow exotic subatomic particles to momentarily wink into existence – giving physicists a brief opportunity to study their behavior.

The LHC is the latest and greatest of these particle colliders. Famously, in 2012, it confirmed the existence of the Higgs boson, a subatomic particle that physicists had first predicted to exist back in the 1960s. Confirmation of its existence filled an important missing piece in the physicists' "Standard Model," a theory that explains how all the known subatomic particles relate to the forces that govern the known Universe.

But physicists know full well that the Standard Model is incomplete. For one thing, it's abundantly clear from studies of the Universe that the known particles can account for no more than 15 percent of matter. The other 85 percent is mysterious "dark matter." There simply have to be dark matter particles – and potentially all manner of other particles – still waiting to be discovered.

Physicists have made all sorts of predictions of where and how these missing subatomic particles might be found. The trouble is, experiments like the LHC – which physicists rely on to prove which of their theories are correct – seem to have hit a wall. They are finding it harder to discover evidence of the new particles that could confirm theoretical predictions, as the story of the "particle" that disappeared in 2016 illustrates.

The LHC has years left to operate. It may still make exciting new discoveries. But, as of early 2018, it's at least possible that the remaining undiscovered particles only pop into existence in extremely high-energy conditions – beyond anything even the LHC can create.

All is not entirely lost. Physicists have evidence that

The Higgs mechanism explains how objects get their mass, with help from the famous Higgs boson.

(Above) The Hadron Collider probes the properties of subatomic particles called neutrinos that could lead to new discoveries in physics.

some of the subatomic particles they already know about behave in ways that don't quite match expectations.

Strange particles called neutrinos were predicted to behave like ghosts: to be visible, but to carry no substance – to be "massless." Then, in the late 1990s, it emerged that neutrinos do carry a tiny amount of substance after all. They can change their state in subtle ways as they fly through space, which is only possible if neutrinos carry a small amount of mass.

Neutrino experiments could help physicists work out how to extend their Standard Model. They could even explain the identity of some forms of dark matter. But, as of 2017, they haven't done so yet.

GO WIDER

SIX DEGREES OF SCIENCE

For more on...
The birth of the Universe:

 The first stars in the Universe

Unexpected physics discoveries:

Have we broken the light barrier?

The mystery at the core of the Universe

HOW TO HIDE INFORMATION IN THE FABRIC OF TIME

It has the makings of the perfect security system. Physicists have worked out a way to effectively pull time apart, creating a gap into which they can insert a secret message. Then, the scientists stitch time back together again. An outside observer is left clueless – unable to see a message that is now hidden inside the very fabric of time.

This is the mind-bending theory behind "time cloaking." And, in 2012, the theory became reality.

Time cloaking is really about playing with the speed of light as much as it is about manipulating time itself. In the vacuum of space, light always travels at the same speed: about one billion kilometers (670 million miles) per hour. Nothing in the Universe moves faster.

But when it's trapped inside optical fibers, light moves more slowly – generally about 0.7 billion kilometers (435 million miles) per hour.

Key to time cloaking is the fact that scientists have developed technology that can manipulate the speed of light in optical fibers. They can interfere with light's individual particles – or photons – so they move through the optical fiber slightly slower than the 0.7 billion kilometers (435 million miles) per hour speed.

You can think of those photons as a little like cars traveling nose-to-tail down a highway. To begin with they are all moving at the same speed – say 60 kilometers (37 miles) per hour. Then the scientists step in with their technology: they slow down some of the cars to 40 kilometers (25 miles) per hour, which means a gap gradually opens up to the cars in front. Eventually, the gap is wide enough that an event can occur inside it. A pedestrian can dash across the road between the speeding cars, for instance.

At this point, the scientists step in again. They stop slowing down the cars behind the gap and focus instead on slowing down the cars in front of the gap. This means the gap begins to narrow. When it disappears altogether, the

There's now a way to effectively pull time apart to conceal data.

scientists switch off their technology – their job is done.

An observer standing further down the road is left scratching their head in amazement, wondering how a pedestrian could possibly have crossed the road between cars that are all traveling nose-to-tail at high speed.

Martin McCall at Imperial College London and his colleagues dreamed up the idea of time cloaks in 2010. The first practical demonstration of the idea – using light in optical fibers, rather than cars on a highway – came in 2012. The gap scientists opened up was so brief as to be impractical. But the following year, an updated version of the technology opened up a gap in the stream of light large enough to conceal a usable chunk of information – an email message, for instance.

If the gap is closed after the message has been inserted, an observer further along the optical fiber won't be able to see it. But if someone else even further down the optical fiber opened up the gap again, the message would become readable once more.

A demonstration of this reopening of time to retrieve a hidden message came in 2014.

In a modern world built around the transfer of sensitive information down optical fibers, it's easy to see how time cloaks could prove useful. In 2015, British insurance company Lloyd's estimated that cybercrime might cost businesses $400 billion – about £330 billion – each year.

The Internet is built on a network of optical fibers, which are ideal for time cloaking.

By the end of the decade, those costs might have risen to $2 trillion (£1.6 trillion). Any technology that can help keep data safe from hackers is sure to be welcome.

Time cloaks aren't the only way that physics can help with online security. Some of the strange quantum properties of subatomic particles could find a role in making networks more secure, too – an idea that moved one step closer to reality with the launch of the first quantum communications satellite in 2016.

G O   W I D E R

SIX DEGREES OF SCIENCE

For more on...
Cloaking technology:
The real-life invisibility cloaks

Manipulating photons of light:
Revealed: why light is so weird

Quantum communications:
The unhackable Internet of the future

HAVE WE BROKEN THE LIGHT BARRIER?

Nothing travels faster than light. Orbiting planets don't. Speeding comets don't. Even the fastest spacecraft currently in development will be capable of traveling at only one-fifth of light speed. Put simply, the speed of light is a fundamental barrier. This is an idea that lies at the very core of Einstein's theories of relativity.

However, following a sensational announcement in September 2011, it briefly seemed like the idea was wrong.

Scientists at the Gran Sasso National Laboratory in Italy were in the middle of a physics experiment. They were testing whether a particular type of subatomic particle – a class of particles called "neutrinos" – can change identity as they whizz along.

To perform their experiments, the physicists in Italy had teamed up with another group of physicists at CERN, a physics lab 730 kilometers (454 miles) to the northwest, on the Swiss-French border. CERN is home to the famous Large Hadron Collider. It's a facility that has no problem generating neutrinos.

Neutrinos hardly ever react with solid matter, so a beam of neutrinos generated at CERN can easily pass through the hundreds of kilometers of solid rock separating the subterranean CERN facility and the Italian laboratory.

The process worked well. A little too well. The neutrinos from CERN arrived at the Italian lab about 60 billionths of a second earlier than anyone expected they would. It was a tiny discrepancy, but a very significant one. It suggested the neutrinos had traveled 0.002 percent faster than the speed of light.

The physicists at the Gran Sasso lab searched and searched for a possible source of error that could explain the discrepancy, but without luck. In September 2011, they made their "faster than light" result public in the hope that physicists elsewhere in the world would be able to make sense of it.

Most of those physicists were convinced the result must be wrong. But surprisingly, many of them hoped it wasn't.

Einstein's ideas are built around the fundamental principle that nothing travels faster than light.

Strange as it might seem, scientists often become most animated at the prospect of disproving predictions of famous and successful scientific theories. And scientific theories don't come much more famous or more successful than Einstein's theories of relativity, which have proved time and again to correctly predict features of the Universe.

As recently as 2016, when physicists announced they had detected gravitational waves, they were confirming a prediction made by Einstein's theories.

Back in 2011, some physicists pointed out that faster-than-light travel might actually be possible. They said the speed of light isn't necessarily a fundamental limit, more a barrier that can't be crossed. It's just about theoretically possible that a strange class of subatomic particles could be stuck on the other side of the barrier, fated to always travel faster than the speed of light.

Scientists even have a name for these hypothetical particles: tachyons. If tachyons really exist, they open a Pandora's Box of weird phenomena, including the possibility of traveling backward in time – something else that should be impossible in Einstein's grand vision of the cosmos.

Physicists at the Gran Sasso National Laboratory in Italy were left scratching their heads in 2011.

Despite these troubling side effects, some physicists were at least willing to entertain the possibility that neutrinos are actually tachyons.

In the end, careful analysis showed they aren't. By March 2012, it was becoming clear that there were inaccuracies in the experiment's timing system. An optical fiber in the apparatus hadn't been attached properly, and there were problems with one of the accurate clocks used to time the experiment. Once the errors had been accounted for, the neutrinos turned out to be flying along at their expected speed after all.

In a sense, the physicists had saved their science – defeating a high-profile challenge to their core theoretical assumptions. But in another sense, the result was a little disappointing. A door to a new world of strange physics had opened in 2011. Confirmation that neutrinos don't disobey Einstein after all slammed shut that door.

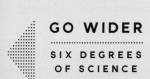

GO WIDER

SIX DEGREES
OF SCIENCE

For more on...
The fastest spacecraft in development:

 The tiny spacecraft with big ambitions

 The mystery at the core of the Universe

The weird science of neutrinos:

 Does particle physics have a problem?

Einstein's predictive powers:

 The waves that distort our planet

HOW THE ATOM BOMB HELPED SAVE THE ELEPHANT

The nature of warfare changed forever in 1945. In the years following the atomic bombings of Hiroshima and Nagasaki, a handful of the world's most powerful nations scrambled to develop nuclear weapons of their own, and improve – if that's the right word – the technology behind them.

The Partial Test Ban Treaty, signed in 1963, restricted any further weapons tests to underground environments, but by then detonations at ground level had already left a lasting legacy. In the 1950s and 1960s, the quantity of radioactive carbon-14 in Earth's atmosphere spiked.

Surprisingly, the surge in atmospheric radioactivity has proved to be of immense value to science. It has helped unravel mysteries of the human brain, crack difficult police cases, and even thwart the poachers who kill endangered elephants for their ivory.

The atmospheric change brought about by atomic tests was dramatic, and it occurred all around the world. These twin facts have encouraged some geologists to use the surge in atmospheric radioactivity to mark the moment that our world entered a brand new epoch. The proposed "Anthropocene" is a chunk of geological time that is characterized by human-driven changes to the natural world – changes that include deforestation, pollution, species extinction and climate change.

Of course, none of those changes have been particularly good for the natural world. And similarly, it's difficult to imagine that flooding the atmosphere with radioactivity was anything other than a bad thing.

However, scientific research has actually benefited enormously from the atomic test radioactivity.

It's all to do with where the radioactive carbon-14 ends up. Plants absorb it when they photosynthesize, and

Atom bomb tests in the middle of the twentieth century flooded the atmosphere with radioactive carbon.

Scientific tools to distinguish legal from illegal ivory could make a difference to the future of the elephant.

then it passes up the food chain into animals, including humans. It gets incorporated into our very DNA.

This is a key point. The amount of carbon-14 in a particular strand of DNA reflects the level of atmospheric carbon-14 when that DNA molecule was being built. Those atmospheric levels have been falling steadily because of nuclear test ban treaties, so every year since the 1950s has a unique atmospheric carbon-14 level.

All scientists have to do is measure the level of carbon-14 in a sample of DNA, match it to the year in which the atmospheric level was at that level, and they know exactly when – between the 1950s and today – the DNA was built.

There are a variety of applications that can benefit from this sort of DNA dating. For forensics teams, it offers a way to work out the age of an unidentified body, speeding up the process of establishing the person's name so that the investigation into their death can proceed.

It is also very useful in the battle against ivory poaching. A ban on the international trade of ivory came into effect in the early 1990s, but, technically, ivory that was traded before then is legal. DNA dating using the level of radioactive carbon can help establish whether a given piece of ivory came from an animal that was alive before or after the trade ban began.

And in studies of the human brain, the carbon-14 from nuclear bombs has proved its worth yet again. It helped establish that the cells in some parts of the human brain are renewed throughout life, providing evidence that the adult brain is much more versatile and adaptable than neuroscientists once thought.

There's a catch, though. Because the radioactive carbon-14 spike is ebbing away year after year, it will soon fall back to its pre-atomic age level. It will stabilize. At that point, each year will no longer have a unique carbon-14 level, and scientists will lose a tool for dating biological tissue samples.

Some hope that new DNA dating methods will emerge that can replace the radioactive carbon-14 clock, but none have yet.

All of this leads to a curious situation. In a matter of years, the long radioactive shadow cast by atomic bomb tests in the mid-twentieth century will be lifted. The natural assumption is that the world will celebrate the moment that happens. But for scientists the world over, the event will make their work much harder.

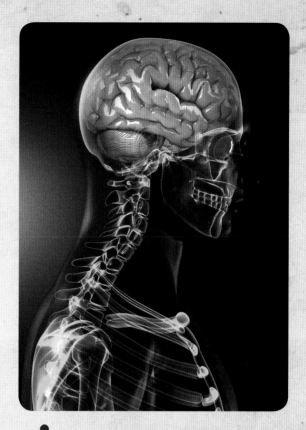

The human brain gains new cells throughout life – a discovery that surprised many neuroscientists.

GO WIDER

SIX DEGREES OF SCIENCE

For more on...
Adapting to life in the Anthropocene:

- Are we living through the Anthropocene?
- Has Chernobyl become a haven for wildlife?

Forensic science:

- The genes that light up after death
- The murder that will never be solved

Brain cell renewal:

- The people with a large chunk of brain missing
- The truth about brain training

THE MAGMA THAT COULD KILL US ALL

They lie asleep, unchanged by the passage of time: two vast lumps of magma deep inside the Earth that seem to have survived intact since the birth of our planet. Once every few tens of million years, one of them wakes up, sending a plume of super-hot material upwards. When it reaches Earth's surface, it erupts in a burst of volcanic activity on a truly awesome scale.

One such episode of volcanic activity probably helped kill off the large dinosaurs 66 million years ago. Another could one day do the same thing to our species.

Scientists know that the Earth is billions of years old, but they also know that it's a very active planet. The interior of the Earth may be difficult to study directly, but geologists suspect it is constantly churning away in a manner that alters its chemistry – so that lava erupting at the surface generally seems chemically "young."

In 2010, scientists stumbled across an anomaly: lava in Greenland that, when it erupted 60 million years ago, was already very, very old. So old, in fact, that some of the chemical isotopes it contains seemed to date back to Earth's birth. The controversial suggestion was that somewhere, deep inside the Earth, must be blobs of hot magma that had escaped the churning process for that entire length of time.

Evidence has now emerged that the blobs have a sinister side.

Life on Earth suffers an occasional devastating mass extinction that wipes out most of the species alive at the time. Many of these extinctions coincide with relatively short-lived periods of intense volcanic activity on a scale that is difficult to imagine. This volcanic activity can blanket millions of square kilometers – equivalent to all of western Europe, for instance – with hot lava. Geologists call these large expanses of lava "large igneous provinces."

The most devastating mass extinction of the last 500 million years – the "end-Permian extinction" – coincided with the formation of one of these large igneous provinces. Most of Siberia was smothered with lava during the end-Permian event.

Even the most famous extinction of them all might have been triggered in part by the formation of a large igneous

Did magma help kill off dinosaurs such as *Tyrannosaurus rex*?

GO WIDER

SIX DEGREES OF SCIENCE

province. The "end-Cretaceous extinction" event about 66 million years ago killed off all the dinosaurs except for one group – the birds. Although there is good evidence that an asteroid slammed into Earth just as the dinosaurs went extinct, there is also strong evidence of a vast outpouring of lava in west central India. It left lava deposits that are 2 kilometers (1.2 miles) thick in places.

What two geologists – Matthew Jackson at Boston University and Richard Carlson at the Carnegie Institution of Washington – realized in 2011 is that the lavas in many of these large igneous provinces are chemically similar. And they seemed to contain the same ancient chemical isotopes associated with the blobs of unmixed hot magma found deep inside the Earth.

The two geologists think that every so often – and for completely mysterious reasons – the blobs wake up and generate a devastating large igneous province with the potential to contribute to a mass extinction of life.

The idea is still very controversial – but if it proves correct, it could be very bad news for the fate of human civilization.

That's because some geologists think the magma blobs still exist. Seismic surveys of Earth's internal structure have revealed two large regions of the lower mantle that seem to be different from the rest of the mantle. One massive blob lies 2,800 kilometers (1,740 miles) below Africa, while its twin lies at the same depth beneath the Pacific Ocean. The suggestion is that these two blobs are the source of the unmixed ancient material that feeds the devastating large igneous provinces.

If – or when – one of the two blobs wakes up again, the large igneous province it forms might well trigger another mass extinction of life on Earth.

A vast outpouring of lava could trigger extinction on a massive scale.

THE MYSTERY AT THE CORE OF THE UNIVERSE

One thousand quadrillion vigintillion. That's a "one" followed by 80 "zeros" – the estimated number of atoms in the known Universe. And just three-dozen subatomic particles detected in Japan could help to explain the existence of them all.

Matter – all of the atoms in the Universe – was created in the Big Bang, billions of years ago. But why so much of it stuck around is still a bit of a mystery. Physicists predict that an equal volume of exotic antimatter should have been produced in the Big Bang, too – and when matter and antimatter meet they are both destroyed, leaving behind nothing but radiation.

Why, then, are there still so many atoms in the Universe? Why are there stars? Planets? Why are there humans?

Where did the antimatter go?

Sometimes what scientists *don't* find is as interesting as what they do find. Rare cases of people who lack large chunks of their brain – but live relatively healthy lives – tell neuroscientists a great deal about the brain's extraordinary ability to adapt. Studies that suggest 90 percent of the human genome serves no useful purpose give evolutionary biologists new insights into the way evolution works.

And discovering that the known Universe appears to contain no appreciable quantities of antimatter tells physicists that something unusual must have happened just after the Big Bang.

One leading idea is that antimatter doesn't behave in quite the same way as matter. Just a tiny mismatch in properties could have helped shift the early Universe away from exactly equal amounts of matter and antimatter, tipping the scales slightly in favor of matter.

In this scenario, matter and antimatter still collided in a mutually self-destructive way, but at the end of that epic cosmic battle, some matter was left standing. Enough to

When matter and antimatter collide, both are destroyed.

fill the known Universe with one thousand quadrillion vigintillion atoms.

It's been decades since physicists first began finding hints of differences in the behavior of matter and antimatter, but those differences were too subtle on their own to explain all of the matter in the Universe.

However, results released in 2016 might just help provide an explanation that can work.

Several physics labs are exploring the properties of strange subatomic particles called neutrinos, which have the curious property of being able to change their identity as they zip along. The scientists create neutrinos in a particle detector and check their identity, then fire them off to another lab hundreds of kilometers away and recheck their identity.

One such experiment made headlines in 2011 for suggesting – incorrectly, as it turned out – that the neutrinos disobeyed Einstein's laws of relativity by making the journey between the two detectors faster than the speed of light.

In 2016, a team of physicists in Japan revealed the latest results from their experiments into identity-swapping neutrinos. Over the course of several years, they had been sending neutrinos – and their antimatter equivalents called antineutrinos – from the Japan Proton Accelerator Research Complex on the country's east coast to the impressive looking Super-Kamiokande detector about 300 kilometers (186 miles) to the west.

In that time, the physicists have spotted 32 neutrinos changing identity in a certain way. Just four of the antineutrinos performed the equivalent identity swap.

Even a non-physicist can appreciate that 32 is a much larger number than four.

Matter and antimatter neutrinos seem to behave differently – a result that could ultimately lead to a satisfactory explanation for the presence of so much matter in the observable Universe today.

But, again as a non-physicist can appreciate, 32 particles plus four particles still equals not very many particles. The 2016 results are promising, but the physicists have far too little data to be sure that they are on to something big. So the work continues.

Galaxies owe their existence to a mysterious mismatch in the properties of matter and antimatter.

Unfortunately, even if the Japanese experiment runs until the mid-2020s, it probably won't gather enough data to provide a rock-solid statistical explanation for the antimatter mystery. And throwing in data from other neutrino experiments now in operation elsewhere in the world might not do the trick either.

In other words, it might be a while before scientists can be sure they understand why the Universe is the way it is.

GO WIDER

SIX DEGREES OF SCIENCE

For more on...
Winding back time to the dawn of the Universe:

 The first stars in the Universe

Does particle physics have a problem?

The neutrinos that seemed to fly faster than light:

Have we broken the light barrier?

The junk in the human genome:

Is 90 percent of our DNA junk?

THE UNHACKABLE INTERNET OF THE FUTURE

Cybercrime is booming. British insurance company Lloyd's estimates the cost it poses to businesses already adds up to $400 billion (£330 billion) each year – a figure that will probably rise in future. As things stand, global financial systems, television stations, state secrets, and even electrical power grids are all vulnerable to attack by cybercriminals.

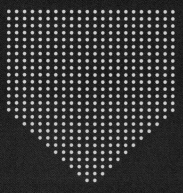

Cybercrime is still on the rise.

It's time to fight back. One satellite launched from China in August 2016 could prove a glimpse of a future in which Internet security is seriously beefed up using the power of science.

That's because this satellite is claimed to offer hack-proof communications. It is the world's first quantum satellite.

Quantum physics emerged in the twentieth century as a new and more accurate way to understand the nature of matter, particularly at the atomic and subatomic scale. But its predictions and conclusions can seem to bear little resemblance to reality as we see it – a point that a physicist called Erwin Schrödinger made in the 1930s. Schrödinger said that quantum physics, which predicts things at the subatomic scale can exist in two states at once, could lead to a situation in which a cat is both alive and dead at the same time.

To most people such a suggestion is clearly nonsense – and quantum physics is, therefore, an arcane irrelevance to daily life. But the $100 million (£82 million) Quantum Experiments at Space Scale (QUESS) satellite might change a few minds. Some physicists think its launch might have fired the starting gun on a quantum space race that could play out over the next several years.

Current technology already offers two people the option of encrypting messages between each other using codes that are, in principle, impossible to crack. But this form of encryption is not tamper-proof: an eavesdropper can listen in when the two people decide exactly which encryption code they will use to protect their messages.

In such a scenario, the eavesdropper will know what the encryption code is, too – and they can then use the information to de-encrypt and read any messages between the two people.

Quantum physics provides a patch for this vulnerability. Physicists can generate pairs of subatomic particles that are intimately linked – entangled – at the quantum level. The link between the two particles is fragile, though: as soon as anyone examines either one of the particles, it changes the entanglement in a detectable way.

This makes entangled particles an ideal tool for two people to use when they make initial contact with each other and decide on an encryption code.

If technology can generate a string of entangled particle pairs, and send one of each pair to person A and the second of each pair to person B, the two people can each use special hardware to guarantee that the communications channel is free from eavesdroppers. It's an idea called quantum key distribution.

The first quantum satellite launched from China in August 2016.

(Opposite) Quantum entanglement promises to make Internet communications unhackable.

Particles of light – called photons – are relatively easy to entangle and send at high speed between the two people. But there is a problem: two entangled photons can't travel very far before the entanglement they share begins to weaken.

This means that ground-to-ground quantum communications links can generally extend over no more than a few hundred kilometers, unless technicians install a great deal of expensive equipment to periodically boost the entanglement signal in a secure way.

This is where the QUESS satellite comes in.

It is orbiting Earth a few hundred kilometers above ground level – near enough that the entangled photons it generates can survive the journey back to Earth and remain entangled. Crucially, the entangled photons don't have to be beamed down to the same point on Earth. They can each be sent to points separated by thousands of kilometers.

The plan is to send one half of a string of entangled photons to a lab in China while the other half of each string is beamed down to a lab in Austria – 7,500 kilometers (4,660 miles) away. If successful, QUESS will instantly expand the range of quantum communications by a factor of 10, turning what has so far been a means of local communication into a global one.

More quantum satellites are already in the pipeline, with launches planned as early as 2019.

As with many revolutionary new communications systems, it's probably governments, military bodies and rich global corporations that will initially feel the benefit of quantum communications – in fact, one bank attempted a local quantum communications channel as long ago as 2004. But the hope is that, with time, even consumers might benefit from the quantum Internet.

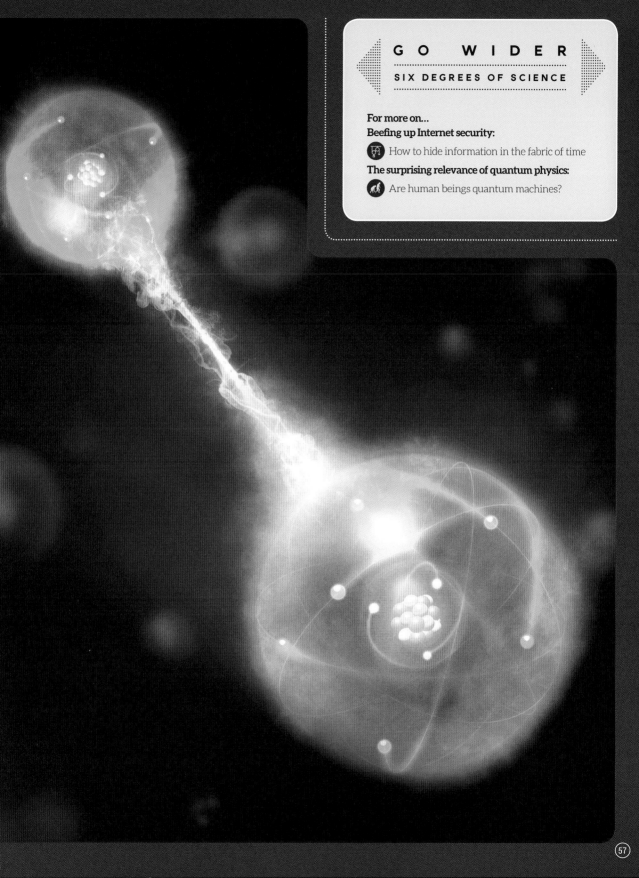

GO WIDER

SIX DEGREES OF SCIENCE

For more on...
Beefing up Internet security:

How to hide information in the fabric of time

The surprising relevance of quantum physics:

Are human beings quantum machines?

REVEALED: WHY LIGHT IS SO WEIRD

Well over a billion images are uploaded to the web every day, but none is quite like the one shared online by physicists on March 2, 2015. In it, they managed to show how light obeys the familiar rules of "classical" physics, while simultaneously following the weird rules of the quantum realm.

This peculiar mix of properties results in a famous concept called wave-particle duality – the idea that light seems to behave in two completely different ways at the same time.

Light puzzled physicists for centuries. Great thinkers like Isaac Newton showed that light behaved as if it were made up of particles – impossibly small, tennis ball-like structures. But experiments by other important physicists such as Thomas Young showed, quite convincingly, that light also seems to behave as a wave, traveling through air rather like water ripples travel across a millpond.

It was only in the twentieth century, with the advent of quantum physics, that physicists faced up to the truth: light is a wave and a particle. And it is both at the same time.

Unfortunately, it is impossible for physicists to actually see light performing this simultaneous double act. A famous concept called Heisenberg's uncertainty principle stands in their way. This puts a practical limit on the ability of physicists to study two physical properties at the subatomic level precisely and at the same time.

Physicists can, for instance, study the way single photons travel along a path as a tiny tennis ball might do. Or they can study the way seas of photons interact with each other to produce complex patterns of waves and ripples. But they can't do both simultaneously.

Light sometimes seems to behave just like a series of waves.

Fabrizio Carbone at the Swiss Federal Institute of Technology in Lausanne and his colleagues haven't measured light behaving as waves and particles at the same time either. But the photograph they took in 2015 does capture features of light's quantum and classical dual state.

Carbone and his colleagues shone bright light at a very short and very thin silver wire. A small amount of that light actually became trapped on the surface of the wire itself, coupling to subatomic particles called electrons in the metal and becoming what physicists call a "surface plasmon polariton."

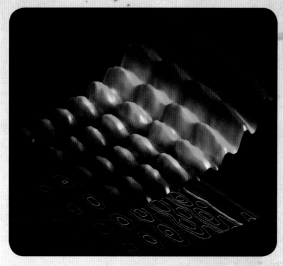

In this form, the light bounced up and down the wire's length like water waves bounce up and down the surface of an aquarium. The light was behaving like a wave.

Next, the scientists fired electrons at the wire. Some of those electrons interacted with the waves of light bouncing up and down the wire, absorbing energy from the light. But the electrons absorbed this light energy in discrete chunks: some electrons gained one "packet" of energy from the light in the wire, some gained two packets, some three – and so on. A few electrons gained as many as eight energy packets.

A simple wave can't hand off discrete packets of energy like that – but quantum objects can, in a particle-like fashion. The fact that the electrons gained discrete energy packets shows that the light in the wire was behaving like a stream of particles.

The photo captures both of these features simultaneously. It's an unusual image, showing what appears to be a broad, multi-colored ribbon rippling from the bottom left of the image up to the top right. To understand the photograph, focus first on the purple ridge in the top right-hand corner. The wire runs along this ridge, from the bottom right to the top left.

Look carefully and you can see four ripples along the purple ridge – these are four wave-like ripples in the light trapped in the wire.

Next, look at the other seven colorful ridges that run parallel to the purple ridge. The wire actually runs along each of these ridges, too – all of the ridges are replicas of the same two-dimensional wire, as "seen" by different

(Above left) Light sometimes seems to act more like a series of ball-like particles.

(Above right) A photograph from 2015 that captures features of light's two states.

electrons that exchanged different amounts of energy with the light in the wire.

The purple ridge shows electrons that gained one energy packet; the dark blue ridge electrons that gained two packets; and so on. Together, these colorful ridges show light's particle properties.

This means the photograph is essentially eight images of the nanowire in one, each showing how the wire "looks" at a different energy level. It's probably apt that the photograph is so odd, though. At the quantum level, light loses its reassuring familiarity and begins to behave in a very weird way.

GO WIDER
SIX DEGREES OF SCIENCE

For more on...
Manipulating light:

 How to hide information in the fabric of time

Quantum physics:

Are human beings quantum machines?

ONE-THIRD OF THE EARTH IS NAMED AFTER THIS MAN

There's a good chance you won't have heard of Percy Bridgman. This twentieth-century American physicist never achieved the fame of Albert Einstein, Niels Bohr or Richard Feynman. But Bridgman has a legacy that no other scientist can match. One-third of the Earth is named after him.

Unfortunately for Bridgman, it's a third none of us will ever see. It's the third of the Earth that makes up our planet's lower mantle – a region of the internal Earth that surrounds the core.

The deep Earth is one of the most mysterious realms known to science, a region of our planet that is very difficult to study directly. Some geologists hope to change that. In December 2015, they began a project to drill through Earth's crust to reach the mantle below, samples of which could lead to a better understanding of how our planet formed 4.54 billion years ago.

Scientists know more about the surface of some distant planets than they do about Earth's deep interior.

But even without those direct samples, scientists have a reasonable idea of Earth's internal structure. By studying the seismic waves that ripple through our planet after earthquakes, they have established that Earth has a dense and solid inner core, surrounded by a liquid outer core, which is in turn surrounded by the mantle, composed largely of solid rock that flows very slowly, like thick toffee might.

Seismic surveys have also helped add some detail to this broad-brush picture of our planet's interior. For instance, they have established that the lower part of the mantle is largely composed of a single mineral, a crystalline substance containing iron, magnesium, silicon and oxygen. This mineral makes up 38 percent of Earth's entire volume.

It is, however, never seen at the Earth's surface, and that poses a problem for geologists. They officially name minerals – quartz, for instance – when they stumble across a physical sample. With no sample of Earth's most abundant mineral, it could not be given an official name.

This changed in 2014 when scientists finally found a sample of the enigmatic mineral. It hadn't been dredged up from the bowels of the Earth – it had actually slammed into Earth in a small meteorite that struck Australia in 1879.

That space rock, the Tenham meteorite, experienced extreme temperatures and pressures as it flew through the Solar System – conditions extreme enough to replicate those found in the lower mantle, which allowed Earth's most common mineral to grow inside the meteorite.

Geologists don't care a great deal about the exact source of a mineral sample. Simply having a real chunk

Bridgmanite

1 mm

Percy Bridgman, the "father of high-pressure experiments."

(Left) The 4.5-billion-year-old Tenham meteorite contains tiny crystals of Earth's most abundant mineral.

of the stuff was enough to name it, even though those individual chunks in the meteorite are just 0.0002 millimeters across.

They gave the mineral its official name – Bridgmanite – in honor of Percy Bridgman. This Nobel-prize-winning physicist is sometimes called the "father of high-pressure experiments" for his work into the way minerals behave under the high-pressure conditions that are found deep inside the Earth.

Bridgmanite joins a list of about 5,000 minerals known to exist on Earth. That's an astonishingly large amount of variety – even given the fact that most are very rare. It's more mineral variety than a planet like Earth should have in normal circumstances.

Many geologists think Earth has such a diversity of rare minerals in part because of the way that living organisms have interacted with the physical world to generate new and unusual crystalline structures. And this means that some minerals act as telltale signatures for the existence of life as we know it.

Some scientists think this fact could prove useful in the search for alien life. The Tenham meteorite shows how space rocks arriving on Earth can provide scientists with samples of Earth's most common minerals. If other meteorites are found to contain samples of some of Earth's least common minerals, they might offer the best evidence yet that alien life exists somewhere else in the cosmos.

GO WIDER

SIX DEGREES OF SCIENCE

For more on...
Mysteries deep inside the Earth:

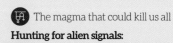

Why Earth's core is younger than you think

The magma that could kill us all

Hunting for alien signals:

The weirdest star in the galaxy

CHAPTER THREE
TECHNOLOGY

THE REAL-LIFE INVISIBILITY CLOAKS

A powerful earthquake strikes a city. Buildings shake violently and crumble to the ground. But amidst this chaos stands a hospital that seems curiously unaffected by the disaster. While other buildings shake, it remains motionless – and when others are flattened, it remains standing.

The hospital is saved because the earthquake doesn't "see" it. The building has been surrounded by a cloak that makes everything inside invisible to the damaging consequences of powerful tremors.

Powerful earthquakes are one of our planet's most destructive forces.

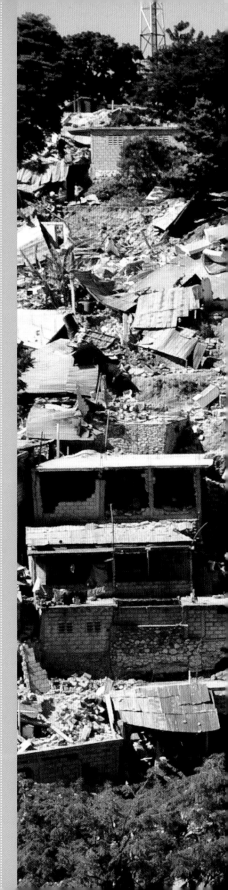

.... Perhaps there is a way to control the destructive power of an earthquake?

The idea of cloaks that make buildings invisible to earthquakes began in the physics lab. Back in 2006, as Harry Potter fans eagerly awaited the seventh and final part of the original series, a team of physicists including John Pendry at Imperial College, London made a sensational announcement.

The scientists said invisibility cloaks – made famous by J. K. Rowling's novels – were theoretically possible to build.

Objects are made visible by the fact that rays of light – traveling in straight lines – bounce off them and reflect into our eyes. The basic concept of the invisibility cloak is straightforward: it "captures" the straight rays of light and bends them in a carefully controlled way so that they are steered around to the far side of an object rather than bouncing off it. Once the light rays have successfully navigated the object, they are released again to continue on the same straight trajectory they had been traveling before they encountered the cloak.

An observer standing near the invisibility cloak is unable to see either the cloak or the object hidden inside.

By the end of 2006, a scientific team led by David Smith at Duke University in Durham, North Carolina, had

actually built a working invisibility cloak that could hide a small cylinder almost perfectly. That is, it could hide the cylinder from microwave radiation, which is – confusingly – a form of light.

In the years that followed, physicists developed invisibility cloaks that work with the visible light our eyes can detect rather than with microwaves, which our eyes cannot see. But, as of early 2018, these "true" invisibility cloaks were generally very small, and tuned to steer light of a very precise color, or wavelength: they can't simultaneously steer all the colors of the rainbow around an object.

In other words, physics is still far from realizing the invisibility cloaks of J. K. Rowling's imagination.

This doesn't mean invisibility cloaking has no value, though. In 2008, Sébastien Guenneau, now at the Institut Fresnel in Marseille, France, and his colleagues began extending the idea in new directions. They realized that invisibility cloak technology could help steer the destructive seismic waves released by earthquakes around human-made structures.

By 2012, Guenneau and his colleagues had begun testing earthquake cloaks in the real world. They – and a few other groups of earthquake cloak researchers – are still trying to work out the physical form that the most effective quake cloak should take. Some have suggested sinking a swarm of large metal rods into the ground that surrounds the building to be cloaked. Others think simply drilling large holes into the ground in a carefully planned pattern could be enough to steer seismic waves around particularly sensitive buildings.

Guenneau and other physicists admit that earthquake cloaks may never be effective enough to completely hide buildings from the effects of the most destructive earthquakes – like the magnitude 9.0 tremor that hit Japan in March 2011. The fear is that the seismic waves from such a powerful quake would destroy the cloak rather than be guided by it. But even if the cloaks are limited to protecting against milder quakes, the technology could still be a game changer.

Meanwhile, the invisibility cloak story continues to march on. Others have developed versions that can hide objects in the stream of time, for instance. And by April 2016, it was the entire planet that two astronomers – David Kipping and Alex Teachey at Columbia University, New York – were proposing to cloak.

They pointed out that astronomers spot alien planets

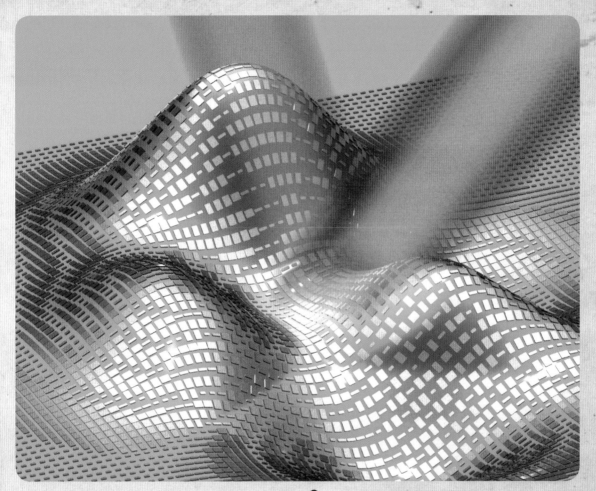

Physicists continue to work on cloaking technology – this "skin cloak" reflects light like a smooth mirror would, even though it is draped over a lumpy object.

orbiting distant stars using the so-called "transit method." When an orbiting exoplanet passes across the face of its star – an event called a transit – it blocks some of the starlight. This leads to periodic patterns of dimming, which a telescope trained on that patch of sky can detect.

If we want to avoid having Earth detected by alien civilizations using the same technique, we could fire powerful lasers into space every time the Earth passes across the face of the Sun to make up for the light that our planet blocks.

Kipping and Teachey point out that alien civilizations might have had the same thought, of course: if there are any super-advanced alien civilizations out there, they might be using this simple kind of invisibility cloak to mask their planet's presence – which might help explain why the known Universe seems disappointingly empty of alien life.

GO WIDER

SIX DEGREES OF SCIENCE

For more on...
Earthquake relief work:

 The drones that control the weather

Time cloaking:

How to hide information in the fabric of time

Super-advanced alien civilizations:

The weirdest star in the galaxy

HOW VIRTUAL REALITY CAN CHANGE LIVES

A foot patrol moves through the quiet streets of an Iraqi city. In an instant everything changes: the bright flash of light is followed almost instantly by a deafening blast – an improvised explosive device.

There are no casualties this time, though: the IED, the people, and even the streets are all computer rendered. This war scene is being played out in virtual reality. And for once, the objective of the computer-generated scene is not to thrill gamers – it's to help former soldiers left traumatized by their experiences of conflict.

Virtual reality has had a long and difficult birth. The technology stretches back decades, but early VR headsets left users complaining of headaches and motion sickness. The best computer processors available at the time weren't fast enough to respond quickly when the wearer turned their head, leading to reports of nausea that crippled the technology.

Today those problems have gone. With the 2015 launch of the Oculus Rift VR headset and the 2016 launch of the HTC Vive, the future is looking brighter for virtual reality. The technology may eventually revolutionize computer gaming, as it has long promised to do.

But virtual reality has dozens of uses beyond entertainment. What makes VR headsets such powerful tools is their ability to put the wearer in someone else's (virtual) shoes. Study after study suggests this sort of embodied experience has a profound impact on behavior that can't be matched using other techniques.

In 2010, for instance, Sriram Kalyanaraman at the University of North Carolina at Chapel Hill and his colleagues asked volunteers to explore a VR simulation of schizophrenia, originally designed to help caregivers understand their patients a little better. A few minutes in the VR environment – experiencing disturbing hallucinations, hearing voices – helped improve the

Battlefield experiences can leave mental scars long after physical injuries have healed.

volunteers' sense of empathy for people in the real world living with schizophrenia.

Grace Ahn at the University of Georgia in Athens (USA) found the same applied for attitudes towards some disabilities. In 2013, she asked volunteers to spend some time controlling a VR avatar with red-green color blindness. After the exercise, the volunteers found it significantly easier to imagine the difficulties of living with the condition. What made the result even more impressive is that, before the VR experience, many of the volunteers described themselves as not being particularly empathetic.

For people living with conditions such as schizophrenia or PTSD, virtual simulations can be just as powerful.

One 2014 study, conducted by Matthew Smith at Northwestern University in Chicago and his colleagues, provides a good example. The researchers asked people with conditions like schizophrenia and bipolar disorder to navigate a VR simulation of a job interview situation – which included help on getting through the ordeal. The volunteers performed much better in real-world job interview–like situations afterward.

And then there are VR simulations for people left with PTSD after experiencing battlefield action. At the University of Southern California in Playa Vista, Albert "Skip" Rizzo and his colleagues have been putting former soldiers back on the virtual streets of Iraq and Afghanistan in a carefully controlled way to help them deal better with their condition.

One of their studies, published in 2010, reported that out of 20 volunteers who completed a course of therapy that included the VR simulations, 16 showed such an improvement in their mental health that they no longer met the official criteria for PTSD.

VR technology may even have the potential to help reverse some of the symptoms of paralysis, if some scientists are to be believed. They have tapped into the brain activity of volunteers who are paralyzed, allowing the volunteers to control an able-bodied virtual avatar through thought alone – an exercise that seems to have helped them regain some use of their limbs in the real world.

There is no doubt that virtual reality has the potential to be a literal game changer in the computer games industry. But it's arguably beyond this commercial realm that the technology will have its biggest impact. For many, virtual reality could be a life changer.

 Immersive virtual environments can change people's attitudes to psychological problems.

GO WIDER
SIX DEGREES OF SCIENCE

For more on...
The medical benefits of virtual reality:

♥ How a man's nose helped him walk again

♥ The first human head transplant

New insights into schizophrenia and PTSD:

🧠 The parasite that may be manipulating your behavior

🧠 Why do we sleep?

🧠 The neuroscientists hunting ghosts

Seeing the world through different eyes:

🧠 The woman with the super vision

THE FUTURE OF THE HAMBURGER

You could almost call it the quarter-of-a-million pounder. At a cost of some £210,000 ($259,000), a hamburger cooked in London in 2013 must rank as the most expensive the world has ever seen. According to tasters, it had an intense taste and a perfect consistency. And it might just represent the future of the meat industry. That's because this burger was grown in a lab, not on the farm.

Humans have been farming livestock for thousands of years, and hunting wild animals for far longer. Eating a nutritious diet based around cooked meat might even have played a pivotal part in the evolution of our large brains, according to some anthropologists.

But feeding today's vast human population on a meat-rich diet is problematic. Put simply, livestock farming is not particularly efficient.

For one thing, the farms can take up huge quantities of land, eating into ecologically and environmentally important habitats. A big chunk of Amazon deforestation stems from a desire for more space for cattle farming. The farms consume vast amounts of water, which puts pressure on authorities trying to maintain adequate drinking water supplies – particularly in world regions that are prone to drought.

And then there are problems associated with livestock itself.

 Demand for new pasture land is driving deforestation across the world.

The world's vast cattle herds generate huge quantities of methane gas when they ferment the grass they munch. Methane is a powerful greenhouse gas, which makes livestock farming a surprisingly important contributory factor to global climate change.

GO WIDER

SIX DEGREES OF SCIENCE

In order to make livestock farms as efficient as possible, it has been standard practice to add antibiotics to animal feed, in order to keep livestock growing at a healthy rate. The emerging problem of antibiotic resistance has arisen in part because of these sorts of practices, which over-expose bacteria to our drugs and give the microbes the opportunity to evolve effective defenses.

Many people are troubled by the ethics involved in eating meat as well.

Artificial meat could avoid almost all of these problems. Some studies suggest a global synthetic meat industry would take up no more than one percent of the land, four percent of the water and produce four percent of the greenhouse gas emissions of livestock farming.

Meat grown in sterile laboratory conditions would not need to be "fed" with antibiotics, either. And since lab-grown meat is not part of a living, breathing animal, lab-grown meat isn't murder: slaughterhouses would no longer be required. Little wonder, then, that animal rights group People for the Ethical Treatment of Animals (PETA) is a strong supporter of the idea.

At Maastricht University in the Netherlands, Mark Post and his colleagues have been working on the idea for some time. It took five years of research – and a large charitable donation from Sergey Brin, co-founder of Google – to turn stem cells taken from cow muscle into the 2013 burger.

Post's team came up with a neat bit of bespoke technology to "exercise" the meat cells as they grew, creating thin strips of pure muscle. Some 20,000 of the strips – grown over several months – provided enough meat for the burger.

In 2016, a new player entered the lab-grown meat market. A San Francisco startup, Memphis Meats, unveiled a synthetic meatball. The cost? About $18,000 (or £14,700) per pound of meat. Clearly production costs will have to fall before lab-grown meat can be a serious commercial prospect.

However, the scientists and entrepreneurs behind these

Lab-grown meat should have none of the environmental or ethical problems of raising animals for slaughter.

efforts are optimistic that the commercial challenges can be met. Post and his colleagues think that their products could be on sale within five years. Uma Valeti, chief executive at Memphis Meats, believes that by the mid-2030s, most meat sold in supermarkets will be lab-grown.

The biggest challenge – one also faced by the genetic modification industry – might be to convince the average consumer to swallow the concept.

For more on...
The problems with farming:

 Crunch time for antibiotics

 Was this humanity's biggest mistake?

Water shortages:

 The dry country that could water the world

Global climate change:

 Are we living through the Anthropocene?

THE END OF THE AUDIO JACK

Technology giant Apple unveiled the iPhone 7 in September 2016, a phone boasting new and improved camera technology, better battery performance and – a first for the famous smart phone – promises of water resistance. But it was one feature the new iPhone lacked that stole all the headlines.

Where was the headphone plug?

In the days that followed, some technology observers made a prediction. Apple's decision to remove the 3.5-millimeter (0.13-inch) audio jack from its iPhones will see the gradual decline and eventual disappearance of the socket from all electronic devices.

Many will mourn its loss, not least because of its long and proud history. In a slightly larger form, the audio jack is a piece of technology that has remained virtually unchanged since Queen Victoria was on the throne.

Appropriately enough, the original audio jack, the quarter-inch jack, was invented largely because of the rise in popularity of the phone – the original telephone.

Shortly after the telephone was invented in 1876, rich homeowners began renting their own. The first phones came in pairs and could only "talk" to each other. Soon, however, people became keen to add to their phone contacts.

In 1878, Alexander Graham Bell – widely considered to be the inventor of the first practical telephone – came up with a solution. His firm, based in Boston, Massachusetts, pioneered the telephone switchboard. Bell hired women telephone operators to answer calls and connect the caller with their desired party.

The quarter-inch (6.35-millimeter) audio jack soon emerged as a quick, easy and reliable way to make the right connections on those early telephone switchboards.

Technology has evolved considerably since the days of the first telephone, but the audio jack has remained almost unchanged. Almost, but not quite. The tips of the very first audio jacks were rounded rather than pointed.

There was just one band of plastic running around the metal shaft rather than the two, three or more seen on some modern jacks. And, of course, they were 0.25 inches (6.35 millimeters) in diameter rather than 3.5 millimeters (0.13 inches).

This most dramatic change to the audio jack occurred in the middle of the twentieth century. With the rise of small, portable radio technology, manufacturers demanded a smaller version of the audio jack. The quarter-inch version didn't disappear entirely, though – it's still popular in the music business, for instance.

The middle of the twentieth century brought other changes. Improvements in technology saw the rise of stereo recordings – another major threat to the survival

The audio jack traces its origins to nineteenth-century telephone switchboards.

of the audio jack. Again, though, it simply evolved. This is when an extra plastic band was added to the jack, separating left and right audio channels, and turning a mono audio plug into a stereo one.

More recently, particularly with the appearance of mobile phones, consumers have become increasingly fond of hands-free operation. The jack evolved again, gaining yet another plastic band to create a third independent audio channel – one that can be hooked up to a microphone built into today's tiny modern earbuds.

But now the audio jack is facing a new threat. This time it might struggle to adapt.

The trend in the twenty-first century is towards ever smaller, ever slimmer and – crucially – ever more versatile electronic devices packed with a rich array of features. Modern smart phones double as digital cameras, Internet browsers, email receivers, GPS devices, games consoles and more.

Apple was looking to expand this versatility concept, and the physical sockets on its iPhones quickly came under the spotlight. Apple decided having a separate charging socket and audio socket no longer made sense. The two are now combined into one versatile socket – which could mean the end of the audio jack.

Some predict that in as little as a decade, the 3.5 millimeter (0.13-inch) audio jack will lose its place as the default headphone connector.

The audio jack rose to popularity because of the telephone. It seems apt that the iPhone, arguably the most iconic of twenty-first-century telephones, should be responsible for its move towards technological extinction.

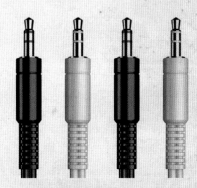

(Top) The iPhone 7 was packed with features, but Apple decided to remove the traditional audio jack.

(Above) Could Apple's decision be the beginning of the end for the longstanding audio jack technology?

◀ G O   W I D E R ▶

SIX DEGREES OF SCIENCE

For more on...
The irresistible lure of smart phones:

Let your car do the driving

The flood story that isn't a myth

Why our brains are shrinking

How headphones shrunk:

The truth about green energy

Einstein's role in GPS:

Why Earth's core is younger than you think

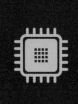

DINOSAUR RESURRECTION: MEET THE CHICKENOSAURS

Some have long, dinosaur-like leg bones. Others have dinosaur-like feet. Yet more have a dinosaur-like snout. And all have grown inside eggs that were laid no more than a few years ago.

Meet the chickenosaurs – the spectacular animals emerging from scientific efforts to understand evolution by reversing it.

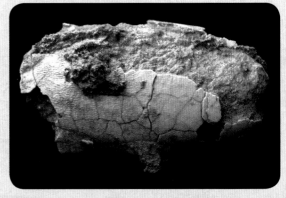

(Left) Modern birds look a lot like prehistoric dinosaurs from the knees down.

(Right) An ancient dinosaur egg has more in common with a modern chicken egg than scientists once thought.

It's common knowledge that all of the large dinosaurs on Earth perished millions of years ago, when their prehistoric ecosystem came crashing down – almost certainly because of a large asteroid and a spell of intense volcanic activity on a scale barely imaginable by modern standards.

What's less widely appreciated is that some dinosaurs survived the catastrophe – and they flourished. These dinosaurs coped so well, in fact, that most people will encounter them every single day. They are the birds.

Birds lived alongside the other dinosaurs for tens of millions of years, until a series of catastrophic events about 66 million years ago. The birds survived. No other dinosaurs did.

Confirmation of the dinosaur-bird link has big implications, both for biologists and for science fiction fans.

Back in 1990, when Michael Crichton wrote the hugely successful novel *Jurassic Park*, he had to devise a mechanism through which dinosaur DNA could survive into the modern world. Crichton's ingenious solution was to imagine that the DNA in dinosaur blood could be preserved in the stomachs of prehistoric biting insects trapped in amber.

Today Crichton's scenario is considered unlikely. DNA can survive for tens of thousands of years, but not millions of years. But with the realization that birds are dinosaurs, biologists no longer need to look for dinosaur DNA in ancient fossils. They can find it in any bird alive today.

And there's more. Birds may be dinosaurs, but they are such a specialized group that their DNA can't on its own tell geneticists a great deal about the other dinosaurs – *Tyrannosaurus* and the like. The scientists need another perspective. They can get that through studying crocodile DNA.

Crocodiles aren't dinosaurs, but the two groups of animals do share a common ancestor. This means that geneticists can get some sense of how particular genes would have looked in large dinosaurs by comparing the way they look in living birds and crocodiles.

Making bird genes look more crocodile-like is, crudely speaking, a little like winding back the clock of evolution: it can result in birds that develop features more like those seen in *Tyrannosaurus* and other large dinosaurs.

Arhat Abzhanov at Harvard University in Cambridge, Massachusetts, is one of a handful of geneticists performing this sort of work. In 2011, his research team subtly altered the genes inside chicken embryos so that they developed *Tyrannosaurus*-like snouts in place of a bird-like beak.

In 2016, João Botelho at the University of Chile in Santiago used some genetic tweaking to grow chicken embryos with long, slim leg bones, more like those of a large prehistoric predatory dinosaur than those of a typical modern bird.

Even understanding the way embryos behave as they develop in the egg can help wind back the evolutionary clock. In 2015, Botelho and his colleagues discovered that the bird's opposable toe – a feature that helps them grasp their perches – begins to develop because bird embryos are so active inside the egg. Simply injecting bird eggs with a drug that reduces embryo activity saw the embryos grow feet without an opposable toe, which made the feet look much more like those of a large prehistoric dinosaur.

For ethical reasons, the scientists behind all of these experiments kill their chickenosaurs before they actually hatch. The purpose of the studies is simply to understand the way that genes influence animal evolution and development.

But Jack Horner, one of the world's most famous paleontologists, is keen to hatch real live chickenosaurs that carry a whole host of large dinosaur features: arms rather than wings, a jaw filled with teeth in place of a beak, and a long, *Tyrannosaurus*-style tail.

Horner thinks it's only a matter of time until such animals walk the Earth. Extinction, he says, no longer has to be permanent.

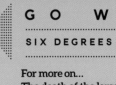

Ancient dinosaurs like *Troodon* have been extinct for millions of years – but animals very like them could walk the world again.

G O   W I D E R

SIX DEGREES OF SCIENCE

For more on…
The death of the large dinosaurs:
- The magma that could kill us all
- The dwarf dinosaurs of Transylvania

DNA's shelf life:
- Encoding the Internet in DNA
- The killers lurking in Earth's ice

Bird behavior:
- Nerd birds love grammar, too!

HOW TO BUILD A STAR ON EARTH

Ours is a one-star Solar System. Earth and the other seven – or possibly eight – planets have been orbiting the Sun for billions of years.

Things might change in the next few decades, though. Our Solar System might be on the verge of gaining a second star. And a third. And a fourth. But these stars won't burn brightly in the darkness of space. They will live right here on Earth. And they might just be the future of the energy industry.

The first commercial nuclear power stations opened in the 1950s. Today they provide about ten percent of the world's electricity. Nuclear power is a proven technology. It's reliable. It's commercially competitive. It produces no climate-warming carbon dioxide.

But nuclear power is also hugely controversial. Very rarely, power stations can go into meltdown, releasing cancer-causing radiation into the environment. Chernobyl, Three Mile Island and Fukushima are, to many people, the three best arguments against continued investment in nuclear power.

More commonly, nuclear power stations generate radioactive waste that no one wants to see buried in their

(Above) The Wendelstein 7-X is at the leading edge of efforts to make nuclear fusion work.

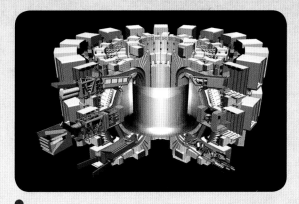

The International Thermonuclear Experimental Reactor has been plagued with problems, but construction is now underway.

local environment. This is where a revolutionary new technology could help. The idea is to generate energy, not by splitting the atom as today's nuclear reactors do, but by fusing atoms together as the Sun does.

Nuclear fusion reactors would take two distinct variants, or isotopes, of hydrogen gas. They would heat the hydrogen to extreme temperatures, turning the gas into something called plasma, in which the electrons are stripped from the atoms to leave the raw atomic nuclei behind. Inside this plasma, the hydrogen nuclei would begin to fuse, releasing energy in the process.

This fusion reaction would – in principle – produce little or no dangerous and long-lived radioactive waste.

Predictably, building an artificial star on Earth is a huge challenge. Scientists have been exploring the idea for decades and they still haven't got a fully working demonstration up and running. It's extraordinarily difficult to generate – and then maintain and control – plasma at the temperature of 100 million degrees Celsius (180 million degrees Fahrenheit) required for fusion to begin.

The best hope is that powerful magnetic fields could corral the plasma, keeping it away from the physical walls of the fusion reactor – which would melt upon contact with anything so hot.

So far, though, efforts to build an experimental system – the International Thermonuclear Experimental Reactor (ITER) – have been hit by setbacks, delays and spiralling costs. Despite these problems, scientists working at the ITER facility in the south of France are optimistic that they are finally on track.

In 2013, construction work began on ITER's "Tokamak" magnetic structure, which would confine the fusion

reaction. The official line, as of 2014, is that the Tokamak will be ready for initial experiments by 2020, and full-on fusion experiments later in the decade. The first commercial fusion reactor based on the ITER technology would follow at a later date.

Others might beat the ITER collaboration to the commercial market, though. In October 2015, work finished on the Wendelstein 7-X – a different type of fusion reactor – located in Greifswald, Germany. By February 2016, the Wendelstein 7-X was producing plasmas from hydrogen gas that reached temperatures of 80 million Celsius (144 million Fahrenheit) – although such plasmas could be controlled for no more than a quarter of a second. The German facility completed its first round of successful experiments a few weeks later, and a year-long upgrade project began later in 2016. Within a few years, its operators hope to be controlling plasmas for 30 minutes.

Scientists in China are making progress, too. In February 2016, there were reports that the Experimental Advanced Superconducting Tokamak (EAST) in Hefei had kept control of plasma at 50 million Celsius (90 million Fahrenheit) for almost two minutes – a record at the time. The EAST operators are ultimately aiming to control plasmas at a temperature of 100 million Celsius (180 million Fahrenheit) for 1,000 seconds, or about 17 minutes.

Even if these efforts succeed in meeting their targets, there's clearly a long way to go before fusion reactors could operate safely for the prolonged periods of time needed for a viable power station. But with several rival groups pursuing their own technologies, it's just possible that the nuclear fusion era is approaching.

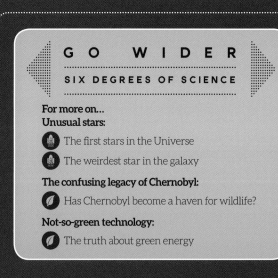

GO WIDER
SIX DEGREES OF SCIENCE

For more on…
Unusual stars:

- The first stars in the Universe
- The weirdest star in the galaxy

The confusing legacy of Chernobyl:

- Has Chernobyl become a haven for wildlife?

Not-so-green technology:

- The truth about green energy

WHAT SYNTHETIC LIFE REVEALS ABOUT THE LIVING WORLD

All it takes is 473 genes. In 2016, biologists designed and created an artificial microbe to explore exactly where the boundary between the living and the non-living world lies.

Their synthetic microbe contains 50 fewer genes than anything found in nature, but it can still grow and reproduce. Its 473 genes are the only ones that are absolutely vital for life – and geneticists don't know what one-third of them actually do.

Biotechnologists are getting more and more sophisticated in their ability to tinker with the internal workings of biological cells. They can add in new genetic machinery from other species to turn microbes into factories that churn out pharmaceutical drugs or food flavorings, for instance.

Since 2008, genomics entrepreneur Craig Venter has had an even grander vision in mind. Rather than manipulating and moving individual genes, he and his colleagues set themselves the challenge of controlling the entire host of genes – the genome – of microbial cells.

By 2010, Venter was able to report an astonishing milestone: he had built an artificial version of the genome of one species of bacteria – *Mycoplasma mycoides* – and then injected his synthetic genome into the hollowed-out cell of a related bacterial species called *Mycoplasma capricolum*.

The strange artificial hybrid powered up and began to move and behave like a normal microbe. Venter named it JCVI-syn1.0.

Venter's work proved hugely controversial outside the scientific community – particularly with those who mistakenly thought that JCVI-syn1.0 was a brand new life form created entirely from scratch. It wasn't: Venter had simply produced a faithful synthetic copy of a perfectly natural genome and shown that it could function normally inside a perfectly natural microbial cell.

Some did appreciate the difference. In its daily

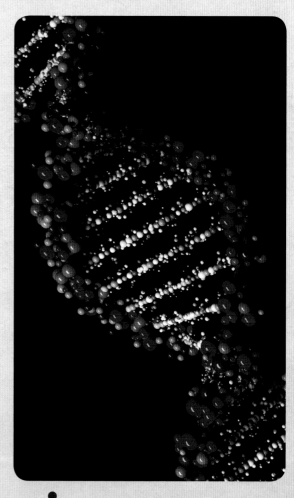

How many genes does life need to function?

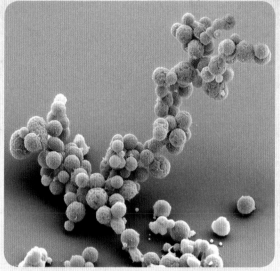

 (Left) Nothing in nature has a genome as small as the synthetic microbe JCVI-syn3.0.

(Above) *Mycoplasma mycoides*, the bacterium that served as a starting point for creating the synthetic microbe.

newspaper, *L'Osservatore Romano*, the Vatican claimed to be cautiously in favor of the work, calling it "high-quality genetic engineering."

Those in the scientific community agreed – although some geneticists questioned exactly what the point of Venter's work was. After all, impressive though JCVI-syn1.0 was, it didn't serve any practical purpose.

Version 3.0 of the microbe, however, did.

By the time Venter unveiled JCVI-syn3.0 to the world in 2016, he had done exhaustive tests to work out which genes he could remove from his synthetic *Mycoplasma mycoides* genome without interfering fatally with its normal functioning. His team removed genes one by one, and tested whether doing so killed the synthetic organism or not.

Venter and his colleagues discovered that there's a lot of stuff in the *Mycoplasma mycoides* genome that the microbe can safely do without – or, at least, that the microbe can do without while it's living in a nice laboratory Petri dish where food is in plentiful supply and rival microbes are absent. Out in the harsh reality of the natural world, some apparently dispensable genes might be required.

But for the *Mycoplasma mycoides* living in the lap of luxury in the lab, Venter and his team found they could whittle down the genome from its 901 genes to just 473.

Those 473 included some genes that play a role in nutrient processing, and some genes that help the genome copy itself when the cell replicates. But, to the astonishment of geneticists the world over, the 473 also included 149 genes with no known purpose.

If the cell is missing just one of these 149 genes, it will die. But, as of early 2018, biologists have no idea why.

This means that Venter's work has presented geneticists with an exciting new challenge. They must discover what those 149 genes do. It's possible that doing so will reveal new insights into the way biological organisms function at the most fundamental level, and refine the distinction between the living and non-living world.

The work should have implications for understanding our biology, too – versions of some of those 149 genes are found in the human genome.

GO WIDER

SIX DEGREES OF SCIENCE

For more on...
Genetic engineering:

 Gene editing just got serious

 How beer could help cure malaria

The surprising human body:

 The secrets hidden inside your body

THE DRONES THAT CONTROL THE WEATHER

On Friday, April 29, 2016, a drone took to the skies above Nevada. Its mission was one that might find favor even with the state's most ardent anti-drone campaigners.

The drone was aiming to bring rain to the drought-prone region.

Nevada sees less rain than almost any other US state.

The idea that scientists – let alone drones – can control the weather and trigger rainfall is hugely controversial. The technology that says it is possible – called "cloud seeding" – has been used for decades, and with many high-profile claims of success. It is said, for instance, that cloud seeding helped control patterns of rainfall during the opening ceremony of the 2008 Beijing Olympic Games, keeping the main stadium dry.

Such claims are difficult to confirm scientifically, though. A 2003 report by the US National Research Council concluded that there is no scientific proof that cloud seeding triggers rainfall – and that finding such proof is probably going to be more challenging than scientists initially thought.

Scientists at the Desert Research Institute (DRI) in Nevada, at least, are convinced that cloud seeding does work. They have been using the technique – with successful results, they say – for more than 30 years.

The science behind cloud seeding is relatively straightforward. Small particles high in the atmosphere can help coax the water molecules in the air into coalescing. Eventually, enough water molecules will congregate around the "seed" particle to form a water droplet that is large enough to fall as rain or snow.

Silver iodide particles are the seeds of choice in many cloud-seeding efforts, because silver iodide has a crystalline structure very similar to that of ice, which makes water molecules more likely to accumulate around the seed.

The DRI scientists have typically fired silver iodide into the atmosphere from the ground – which, clearly, isn't an ideal way of delivering the seeds to the exact location in the atmosphere where they could be most effective. Scattering the crystals into the atmosphere from the air is preferable, but manned flights solely for the purpose of cloud seeding are expensive.

Drones are a way to solve the problem. They are relatively inexpensive – it's even possible for consumers to build a drone from scratch with a 3D printer. Even better, if one were to crash during a cloud-seeding exercise, the risk to human life is very low.

Throughout the first half of 2016, the DRI scientists performed test flights involving their drone – the Sandoval Silver State Seeder. By June 2016, the drone had been equipped with the silver iodide flares it would need to fire up during a cloud-seeding flight.

Drones are proving their worth in all sorts of other scientific contexts, too.

Many people express legitimate concerns about privacy breaches in a world where drones equipped with cameras are becoming more common. But in Africa's National Parks, the sort of discreet snooping that drones offer can be useful: they are a great tool for keeping tabs on poachers and collecting evidence that can lead to convictions.

Silver iodide flares attached to a standard plane – drones equipped with flares are far cheaper and safer.

Drones offer a great way to re-establish contact with remote communities after a devastating earthquake.

Consumers in the US and UK might think that proposals to use drones for package delivery are little more than publicity stunts. However, in impoverished areas of the world where good roads are few and far between, drones are invaluable for delivering medicines to remote communities. They have also proved useful in re-establishing contact between towns and villages in the aftermath of earthquakes or other natural disasters.

It's all too easy to see drones as little more than expensive toys. However, the reality is that drones are aid-delivering, wildlife-protecting – and potentially rain-making – marvels of modern technology.

GO WIDER

SIX DEGREES OF SCIENCE

For more on...
Controlling the natural world:

The real-life invisibility cloaks

Saving the planet one eruption at a time

Battling drought:

The dry country that could water the world

3D Printers:

Human organs on demand

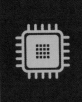

LET YOUR CAR DO THE DRIVING

In August 2016, Ford announced details of a new car it plans to build by 2021. It's going to come equipped with some state-of-the-art equipment: camera and radar sensors, technology to judge the distance from obstacles, and fancy computer software to process all of the information.

But it's what the car will lack that stole all the headlines. There will be no accelerator, no clutch and no brake pedal. There won't even be a steering wheel.

Some people might doubt that the self-driving car revolution is upon us. In truth, it actually started years ago. For instance, mining companies – always exploring how new technology can boost their profits – began using driverless trucks in remote areas of Australia almost a decade ago. Their robo-vehicles have now helped shift millions of tons of material.

There is, of course, a world of difference between driverless trucks operating in and around remote mines in rural Australia, and driverless cars navigating the busy streets of an urban metropolis. Many people are

Some drivers are already prepared to relinquish control of the car to a computer.

legitimately concerned about the safety implications of handing the steering wheel over to an artificial intelligence – or even, as Ford plans, doing away with it altogether.

Volvo has a message for concerned drivers. The carmaker says those who drive its cars already benefit – perhaps unknowingly in some cases – from computer assistance to keep them safe. The Swedish firm is so confident in its computer technology that it has promised there will be no deaths or serious injuries involving new Volvo cars by 2020.

Many cars come with cruise control: Volvo has a more sophisticated version that uses information from a radar system built into the vehicle to adjust the car's speed and maintain a safe distance to the car ahead.

Some new Volvos also use cameras to look for signs that the car is drifting out of its lane. The on-board computer sets off an alert if any drifting is detected so that the driver can take action to return the car to the center of the lane.

And in 2014, Volvo announced it was researching dashboard-mounted sensors to watch the driver for signs of drowsiness – drooping eyelids, nodding heads. The car's other safety features could be programmed to intervene and take control of the car to prevent an accident if the driver is too tired or distracted to take action.

But it's perhaps around the controversial area of the use of mobile phones in cars that computers can do the most to prevent accidents.

Surveys suggest that more than half of drivers on US roads admit to using their phones while driving. Thousands of people are injured or killed on the roads every year as a consequence of this behavior. The distraction risk is particularly severe in the busy traffic that can build up at road intersections.

Some cars already come with technology that can help reduce the problem. These systems automatically block phone calls when the car detects that the driver is changing lanes or turning – manoeuvres that require the driver's full attention. Such systems might already have helped prevent countless accidents.

Even so, there is still an important distinction between computer systems that work with the driver to prevent an accident – as Volvo's systems do – and technology that takes full control of the car at all times. Skeptics who are unconvinced that self-driving cars really are ready for the roads do have a few case studies they can point to in order to justify their concerns. Most significantly, in May 2016 one individual died while operating his Tesla Model S car in "autopilot" mode. His was the first fatality of its kind on US roads. Clearly, self-driving car technology is still a work in progress.

However, the technological world is evolving rapidly. Self-driving car technology is improving – and, at the same time, smart phones and other handheld computing devices are becoming more sophisticated and enticing. We might soon arrive at the point where drivers feel that the technology landscape has shifted such that they are comfortable with handing over control of the car to a computer in order to spend more time using their ever more versatile smart phones.

Google is just one of the companies racing to develop self-driving cars for the consumer market.

GO WIDER

SIX DEGREES OF SCIENCE

For more on…
Mining technology:
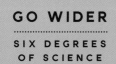 How to get rich in space

Versatile smart phones:
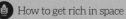 The end of the audio jack

ENCODING THE INTERNET IN DNA

Can you guess which book has been reproduced most often? *The Complete Works of Shakespeare*, perhaps? The *Oxford English Dictionary*? The Bible? In 2012, one book skyrocketed to the top of the list. *Regenesis: How Synthetic Biology Will Reinvent Nature and Ourselves in DNA* by George Church and Ed Regis.

It was Church himself – a geneticist at the Harvard Medical School in Boston, Massachusetts – who reproduced the book. He did so an astonishing 70 billion times. Back in 2012, that would have allowed him to provide everyone on Earth with almost ten copies each. But the world's population would have trouble reading his tome, and not just because of its technical language.

Each copy was encoded, in digital format, in microscopic DNA.

Church and Regis's book – all 53,400 words of it, together with 11 images – takes up about five megabits when converted to digital zeros and ones. DNA's genetic code contains four "letters" – A, T, G and C. Church and his colleagues used two of the letters to represent "zero" and two to represent "one," and then built a DNA molecule from scratch with the correct sequence of letters to reproduce the book's digital format.

The final stage – reproducing the book billions of times – was easy: DNA naturally copies itself.

The exercise confirmed that biotechnology has become sophisticated enough to harness the data-storing power of DNA – and big players in computer technology were watching with interest. In 2016, it was engineers at Microsoft who took the idea of DNA storage further.

(Above right) DNA fingerprinting can identify individuals by features of their DNA – just one way in which the genetic material is used in the modern world.

(Right) The "letters" in DNA sequences are now being used by geneticists to encode data in digital form.

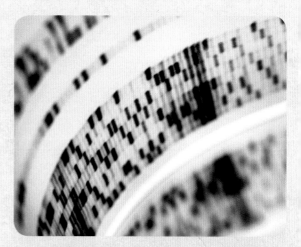

They managed to write about 1,600 megabits of digital data into DNA.

There are good reasons for firms like Microsoft to invest in DNA storage technology. We are churning out information at a faster and faster pace. All those YouTube videos and Instagram photos have to be stored somewhere, and data storage technology is struggling to keep up.

DNA might be the answer, particularly since it takes up very little room. Microsoft estimates that a shoebox full of DNA could potentially store all of the publicly accessible data on the Internet.

And then there's DNA's extraordinary staying power. In the right circumstances – cold and dry conditions – DNA can survive for tens of thousands of years. Geneticists have demonstrated this by pulling DNA from the bones of prehistoric humans – some living about 430,000 years ago – and reading the genetic information it contains. Doing so has revolutionized their understanding of human evolution.

It might even one day be possible to use samples of ancient DNA to resurrect species, such as the woolly mammoth, that were driven to extinction in the last few tens of thousands of years – a real testament to the ability of DNA to endure.

There's one final but important reason why DNA storage makes sense. Because it's the storage system of life itself, DNA will always be current.

Over the past 40 years, we've seen cassette tapes, VHS, floppy disks, CDs, DVDs and more all having their moment in the sun as information storage devices. Each has – or had – its passionate devotees, but each was still eventually consigned to the dustbin of history. And every time technology moves on, piles of data are left behind, fossilized in an old format that is difficult to access.

This shouldn't happen with DNA. As long as there is intelligent life on Earth, there will be interest in reading DNA.

In fact, in the 1970s, two Japanese scientists – Hiromitsu Yokoo and Tairo Oshima – speculated that alien civilizations, if they exist, might be aware of this fact. These extraterrestrials may have left messages for us to decode in the DNA of Earth's life forms, suggested Yokoo and Oshima, perhaps buried in sections of the genome that don't appear to serve any useful purpose.

The pair even went looking for evidence of one of these hypothetical alien messages in a viral genome – although they didn't have much success.

Unlike cassette tapes, DNA will always be relevant to humans – making it a good medium for data storage.

GO WIDER

SIX DEGREES OF SCIENCE

For more on...
Resurrecting extinct species:

Dinosaur resurrection: meet the chickenosaurs

Human DNA:

Extinct humans found in our DNA

Is 90 percent of our DNA junk?

Possible evidence for alien civilizations:

The weirdest star in the galaxy

CHAPTER FOUR

ENVIRONMENT

ARE WE LIVING THROUGH THE ANTHROPOCENE?

It's Millennium Eve, and most parts of the world are taking full advantage of the opportunity to party like it's 1999. People are hours away from seeing the global calendar tick around to the year 2000 – the most significant date change in living memory.

According to some scientific thinking, though, a far more remarkable date may have occurred about 50 years earlier. On, or around, Sunday, January 1, 1950, our planet may have entered a brand new geological epoch: the Anthropocene.

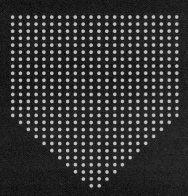

Some landscapes have changed beyond all recognition through human activities.

The Earth is extraordinarily old. It came into existence about 4.54 billion – 4,540,000,000 – years ago. That's a lot of zeroes. To avoid getting muddled, geologists divide that time into chunks and give each chunk a name. It's a lot easier to remember that the dinosaur Diplodocus lived during the Jurassic rather than to recall that it roamed the Earth between 154 and 152 million years ago.

A lot of thought goes into dividing up geological time. Scientists put boundaries where there's evidence that Earth has undergone a major upheaval – a mass extinction of life, or the end of an ice age, for instance. The question those researchers are grappling with right now is a simple one: has human activity changed the planet dramatically and rapidly enough to justify adding a new division to geological time?

Perhaps it has. In the last 50,000 years, our species has conquered all four corners of the world. Humans have had a prominent role in the extinction of many species large and small. We've hacked down forests and turned them into farmland. And we've dug up and burned so much coal and oil that there's near universal scientific agreement that the planet's atmosphere is warming and ocean acidity levels are rising – to the detriment of coral reefs and other marine life.

At a scientific conference in 2016, a team of geologists led by Jan Zalasiewicz at the University of Leicester, UK, and Colin Waters at the British Geological Survey encouraged other scientists to begin debating whether all of this is enough to justify adding the Anthropocene to the geological timescale.

It's an irreversible decision, though, so the researchers will want to be sure that our Earth has permanently changed.

Global temperatures seem to be on an inexorable rise, but in principle they could slow if society makes a concerted move to using power sources that don't involve burning fossil fuels. Science and technology might even offer ways to bring down temperatures to pre-industrial levels.

Deforestation continues apace: it's often the demand for

Slash and burn agriculture destroys vast areas of forest every year.

farmland that's to blame. Efforts to grow meat in the lab rather than taking it from livestock could one day lead to a significant drop in demand for pasture, which might allow forests to reclaim lost territory.

Even the loss of species need not be permanent, according to some bold scientific claims. Some geneticists hope that once-extinct species might be resurrected using DNA extracted from their ancient bones.

If the scientists decide we really are living in the Anthropocene, they'll have lots more to debate – particularly working out when exactly the epoch began.

GO WIDER

SIX DEGREES OF SCIENCE

For more on…

Humanity's impact on the natural world:

 Has Chernobyl become a haven for wildlife?

 The killers lurking in Earth's ice

When life kills itself

Human activities have had an impact on the natural world for at least 10,000 years, when farming began in earnest.

Which brings us back to New Year's Day 1950. This date, or one near it, is the preferred choice of scientists such as Zalasiewicz and Waters. Why pick that quiet Sunday morning to mark the moment that our planet lurched into a new epoch?

Blame the bomb. There were so many atomic weapon tests between 1945 and the early 1960s that levels of

The race to develop the atomic bomb flooded the atmosphere with radioactivity.

radioactivity in the atmosphere spiked. A new geological epoch should be tied to a dramatic event that happened at the same time all around the world. The rise in atmospheric radioactivity as a consequence of atomic testing would make a suitably dramatic choice.

How human activity has affected our bodies:

 Was this humanity's biggest mistake?

How the atom bomb helped save the elephant

Is a 250-million-year-old extinction event killing humans today?

Reducing our impact on the world:

The future of the hamburger

Can super-coral save our seas?

Saving the planet one eruption at a time

SAVING THE PLANET ONE ERUPTION AT A TIME

They call 1816 the year without a summer. A volcanic eruption in Indonesia the previous year had injected colossal quantities of dust into the stratosphere. The dust spread and, by 1816, it was blocking out some of the light from the Sun. Global temperatures dipped half a degree. Crops failed. Food supplies ran low. There was hunger and violence across Europe.

(Above left) Large volcanic eruptions can send plumes of dust high into the atmosphere.

(Above) Marine life is under pressure because of climate change.

It stands as an example of the formidable power volcanoes have to shape the world and cool the climate. And, as global warming becomes more severe, that has got some scientists thinking.

Should we consider artificially simulating volcanic eruptions to try to slow the pace of climate change?

The global climate is warming in an alarming way. The year 2014 was, at the time, the hottest year in recorded history. But then 2015 came along – it was even hotter. 2016 was even hotter than 2015.

Ideally, world leaders would take action to slow the rising global temperature, but many climate scientists fear that there isn't yet the political will to make a real difference. A few scientists think society needs a Plan B – a stopgap

measure to reduce the temperature rise in the short term.

Some of them put their faith in the controversial field of "geoengineering."

The aim of geoengineering is to artificially reproduce natural processes that can help to lower temperatures or reverse other consequences of climate change. One plan is to encourage photosynthetic oceanic plankton to bloom. In theory, the plankton pull carbon dioxide out of the atmosphere and then, when they die and drift to the seafloor, lock the greenhouse gas away where it can cause no harm.

Plankton blooms occur naturally, but some scientists think they can be generated more frequently by fertilizing the ocean with iron, a key plankton nutrient. However, deliberately dumping iron particles into the ocean is very controversial. Many scientists and environmentalists alike fear the iron could have unpredictable effects for the entire ocean ecosystem.

As such, there have been few experiments into oceanic iron fertilization. The idea is not entirely dead in the water, though: one test carried out in the Southern Ocean in 2004 seemed to work more or less as planned, according to a detailed study published in 2012.

Iron is not the only thing geoengineers propose dumping into the ocean. Ground-up rock could help buffer the ocean's pH – which is dropping because carbon dioxide in the atmosphere forms a mild acid when it is absorbed by ocean water. Staving off this "ocean acidification" could prove enormously beneficial for coral reefs and other marine species.

As recently as 2017, the idea of using a common rock mineral called olivine to reduce the effects of ocean acidification was under discussion at international scientific conferences.

Weather balloons, meanwhile, could help "seed" wispy cirrus clouds with tiny chemical particles that encourage the formation of ice crystals. Normally, cirrus clouds trap heat in the atmosphere. But with the right balance of ice crystals, the clouds should become more transparent and let more heat escape into space.

And then there's the idea of mimicking volcanic eruptions. This would involve sending up weather balloons carrying payloads of sulfur-containing particles with a similar composition to volcanic dust. In theory, this form of geoengineering could offset a significant chunk of the recent atmospheric warming – as natural "experiments" like the 1816 eruption show.

Phytoplankton blooms could help remove carbon dioxide from the atmosphere.

But – as 1816 demonstrates again – this form of geoengineering may carry a huge price. It could change rainfall patterns in unpredictable ways, impacting crop growth and leading to food shortages, just as in the year without a summer.

However, climate warming means that food shortages are likely to become more common anyway. Some observers think the world might soon reach a point where geoengineering the planet is the lesser of two evils.

GO WIDER
SIX DEGREES OF SCIENCE

For more on...
Climate change:
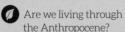 Are we living through the Anthropocene?

Ⓔ The microbes that eat (and poo) electricity

Fixing the problem:
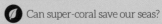 Can super-coral save our seas?

Cloud seeding:
Ⓜ The drones that control the weather

WHEN FOOD BITES BACK

Food has an impact on our bodies – you are, after all, what you eat. But some scientists have discovered that an animal's food can influence its body in a far more surprising, even sinister way.

Food can kill an animal by interfering with its DNA – a discovery that may revolutionize the pest control methods used by crop farmers.

DNA is probably the most famous biological molecule of them all. But it's not the only carrier of genetic information. A related molecule called RNA can store genetic information, too.

In animal cells, RNA often acts as a messenger. DNA is essentially a library of information – that information is put to practical use in factory-like structures elsewhere in the cell, where the genetic code is "read" and used to build things. "Messenger" RNA carries the genetic information between the library and the factory.

RNA has a different role in some viruses. Here, it is the primary carrier of genetic information, which means that these viruses have a genome – their library – encoded in RNA rather than DNA.

RNA viruses can invade and destroy animal cells. Very early in their evolution, though, animals developed a primitive immune system that could respond to the

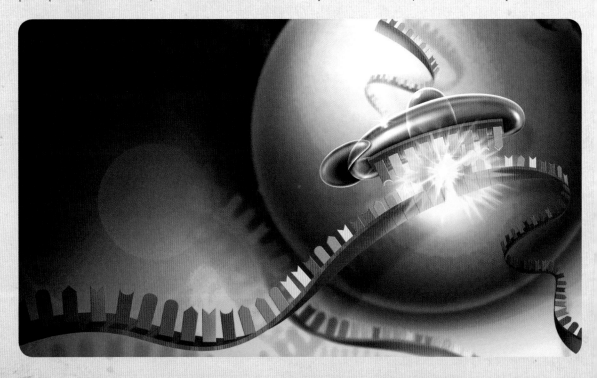

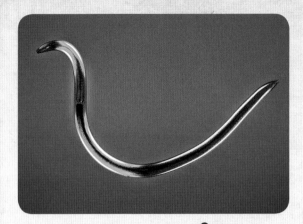

Nematode worms can be genetically manipulated by the food they eat.

(Opposite) RNA interference is an ancient tool that organisms use to destroy messenger RNA.

Maryland, first demonstrated this process in 1998. They genetically modified bacteria to produce RNA that could be confused for the RNA in the cells of microscopic worms. Then they let those worms feast on the bacteria. After the meal, the worms began to twitch in a way that suggested their cells were malfunctioning.

Other biologists have taken this idea much further. They have genetically modified crop plants to produce RNA identical to the messenger RNA inside the cells of troublesome insect pests. Now, shortly after those pests begin to nibble on the crop, their cells begin to malfunction and the insects die. Best of all, the plant shouldn't kill other insects that live in the fields but that don't eat the crop – so insect biodiversity remains high.

Such plants could transform agriculture. Agrochemical giant Monsanto hopes farmers will be growing corn that exploits this vulnerability by the end of the decade. Other firms around the world are developing crop sprays that exploit the same vulnerability.

The big question still being debated is whether food has the power to manipulate human DNA. Reassuringly, most scientists think it can't. Numerous scientific studies suggest that only simple animals like worms and insects can be controlled in this way: there's no convincing evidence that food can fatally manipulate the genes of complex animals like mammals. Even so, the fact that this strange phenomenon exists at all in the animal kingdom is certainly food for thought.

threat. Their cells carefully monitor the level of RNA they contain. If the level rises too high, which might hint at the presence of an RNA virus, the immune system destroys the excess RNA to neutralize the invader.

There's just one problem with this approach. The primitive immune system can't always tell the difference between viral RNA and the messenger RNA that the cell itself generates. Sometimes it can get confused and attack the wrong RNA. That can be a big problem for the animal.

If its primitive immune system targets and destroys messenger RNA, the cell loses the vital line of communication between its DNA library and its cellular factories. Production grinds to a halt, and the cell begins to malfunction. It's a phenomenon scientists have termed "RNA interference."

Biologists studying RNA interference soon realized that it means some animals have a bizarre vulnerability: the food they eat can, in principle, cause their cells to malfunction.

Here's how. An animal's body digests food, and its cells then absorb nutrients from the meal. Supposing those nutrients include bits of RNA that happen to look just like the animal's own messenger RNA. The cell's immune system assumes the influx is a sign of a virus and begins attacking any RNA that looks like these invading molecules – which means it begins destroying the cell's own messenger RNA.

Biologists Lisa Timmons and Andrew Fire, then at the Carnegie Institution of Washington in Baltimore,

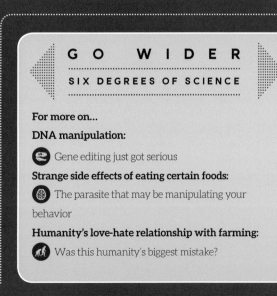

G O   W I D E R

SIX DEGREES OF SCIENCE

For more on...

DNA manipulation:

Gene editing just got serious

Strange side effects of eating certain foods:

The parasite that may be manipulating your behavior

Humanity's love-hate relationship with farming:

Was this humanity's biggest mistake?

THE DRY COUNTRY THAT COULD WATER THE WORLD

Israel, a country in an arid region with a decade-long history of severe drought, has a water problem. But it's not the one you might expect. What should the country do with all of the surplus drinking water it has sloshing about?

Early in 2014, Israel was experiencing its lowest rainfall in 150 years. Jerusalem reportedly saw next to no rain at all in January. In those circumstances, water shortages are more or less inevitable. Not in Israel: the country's 2014 water supply met all of its demands – and then some.

Israel's secret? Extremely efficient water recycling certainly helps: the country recycles and reuses more than 80 percent of its wastewater in agriculture and industry. Strict monitoring of the nation's plumbing is important, too, which means Israel loses only about ten percent of its water supply to leaks – in some western countries the figure is closer to 40 percent.

But the real key is "desalination" plants that can turn salty Mediterranean seawater into cool, clean, fresh drinking water.

Desalination is not a new technology. The US opened its first desalination plant in the early 1960s and continues to open new plants to this day. But it's expensive. Estimates suggest desalination plants in California provide freshwater from the Pacific Ocean at prices of around $2.50 (about £2.00) per 1,000 liters (220 gallons).

In Israel, the figure is reportedly down to roughly 58 cents (46p) for the same quantity of water.

The country has managed to lower costs through careful scientific research into desalination technology. Water researcher Edo Bar-Zeev at the Zuckerberg Institute for Water Research in Negev, Israel, has helped with those efforts.

In 2013, for example, Bar-Zeev and his colleagues took a closer look at the process that desalination companies use to remove particulates from seawater during pre-

Irrigation systems in Israel make extensive use of recycled wastewater.

treatment before it passes into the desalination plant.

Typically, desalination firms add special chemicals that cause particulates to coagulate for easier removal. But chemicals cost money – and there is always the potential for those chemicals to flush back into the Mediterranean Sea where they might pose a risk to wildlife.

Bar-Zeev and his colleagues wondered whether a cheaper, chemical-free biological filtration method already used to remove particulates from wastewater could also be used to pre-treat seawater before it is stripped of its salt.

The assumption was that it couldn't. Wastewater trickles very slowly through biological filtration systems. The water flowing into a desalination plant passes through the filtration system quickly – too quickly, scientists thought, for microbes to remove the particulates it contains.

However, Bar-Zeev and his colleagues found that this assumption was incorrect. They conducted a 12-month-

long trial, and found that a microbial-based filtration system was just as effective at removing particulates from seawater as a standard chemical filtration system. The microbial filter is much cheaper, and potentially more environmentally friendly.

Technological breakthroughs like this help explain why Israel has managed to lower the cost of desalination to a competitive level.

Israel's desalination plants are helping the country exceed its water requirements.

The technology could even help reduce political and social tensions in the eastern Mediterranean. There are hopes of international conferences in future that will bring together water scientists from several countries across the region to discuss and share desalination technologies.

GO WIDER
SIX DEGREES
OF SCIENCE

For more on...
Dealing with water shortages:

How to get rich in space

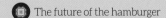

The future of the hamburger

The drones that control the weather

IS A 250-MILLION-YEAR-OLD EXTINCTION EVENT KILLING HUMANS TODAY?

About a quarter of a billion years ago, the mother of all catastrophes hit animal and plant life. Almost 90 percent of species were wiped out in the biggest mass extinction of the last 500 million years.

In 2009, geologists made an astonishing announcement. The extinction, they say, is still claiming lives today – human lives.

There is a curious epidemic of lung cancer in Xuanwei, a county in China's Yunnan province. Cancer cases there are up to 20 times higher than in other parts of China – and the disease affects smokers and non-smokers alike.

Scientific research suggests the culprit is the coal that the locals use for heating and cooking. Burning coal releases molecules called polycyclic aromatic hydrocarbons (PAHs), which are known to be potent carcinogens.

However, some scientists think PAHs can't be the full explanation – particularly since coal is used for domestic heating and cooking elsewhere in China, where the incidence of lung cancer is lower. Is there something unusual about the coal in Xuanwei County that could explain why it seems to be particularly dangerous?

Geologist David Large at the University of Nottingham, UK, and his colleagues believe there is.

They say that the coal in Xuanwei County is unusually rich in silica (essentially fine-grained sand). It is also the only coal used domestically in China that formed during those tumultuous times 250 million years ago – an event known as the end-Permian mass extinction. Large and his team think these two facts are linked.

Exactly what triggered the devastating end-Permian mass extinction event is still being debated, but it almost certainly involved volcanoes – a lot of volcanoes. At exactly the same time as the extinction was unfolding, volcanic activity blanketed most of Siberia with hot lava.

Large and his colleagues say that vast quantities of volcanic gas were belched out with the lava. These gases almost certainly generated strongly acidic rain,

The 2-meter- (6-foot-) long *Dinogorgon* was wiped out in the end-Permian mass extinction.

 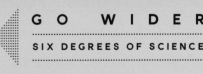
Volcanic activity may have made one coal deposit unusually harmful to human health.

which ate away at any exposed bedrock – in the same way that mild acid rain in the recent past has dissolved stone architecture in some cities.

As a consequence of this extreme acid rain 250 million years ago, the groundwater quickly became rich in silica – tiny grains of sand that came from dissolving the bedrock.

Some of this silica-rich groundwater flowed beneath the area that is now Xuanwei County. At the time, this region was covered in lush, humid forests. Such forests generate peaty soils that can eventually turn into coal. But before this peat crystallized into coal, the silica-rich groundwater seeped in. This explains why the coal in Xuanwei contains so much silica.

When the locals burn the coal, they inadvertently release that silica into the air. Large and his colleagues speculate that it somehow makes the coal fumes even more carcinogenic.

Maybe that's because silica itself can trigger cancer – the International Agency for Research on Cancer has classified silica as carcinogenic to humans. Alternatively, silica might somehow help deliver the PAHs into human lungs more effectively to make them more dangerous to human health.

Not all cancer researchers are convinced that the silica in the Xuanwei coal can make that much difference. They think there might be other dangerous chemicals in the coal – although they do recommend more thorough scientific assessments to work out how much damage the silica in the coal could do.

Clearly, there is still plenty of work to be done before Large and his colleagues can confirm their idea, but their hypothesis is certainly plausible. It's possible that the calamitous environmental events that occurred 250 million years ago have reached through time to devastating effect.

GO WIDER
SIX DEGREES OF SCIENCE

For more on...
The link between volcanoes and extinctions:
The magma that could kill us all

Humanity's problems with coal:
Are we living through the Anthropocene?

The roots of cancer:
Why cancer is like a selfish animal

HAS CHERNOBYL BECOME A HAVEN FOR WILDLIFE?

Early on the morning of Saturday, April 26, 1986, technicians at the Chernobyl nuclear power station in what is now Ukraine began testing the safety of the reactor. The tests didn't go as planned. Far from demonstrating the safety of the plant, they triggered a chain of events that ended in nuclear meltdown.

More than 30 years on from what was arguably the worst nuclear power plant accident in history, something unexpected has occurred. Some biologists have begun reporting signs that the wildlife around Chernobyl is actually thriving.

Nuclear disasters, they say, are not as environmentally and ecologically catastrophic as most people might assume.

In the aftermath of the Chernobyl accident, the authorities took no chances. With plumes of radioactive material billowing up into the atmosphere, they quickly decided to pull hundreds of thousands of people from the surrounding area, forcing them to abandon their homes in order to protect their health.

This was undoubtedly the right call. The immediate environmental effects of the nuclear meltdown were severe. Within months, though, nature apparently began to bounce back.

By the late 1980s, aerial surveys suggested deer and wild boar populations in the Chernobyl exclusion zone were on the rise. Smaller mammals seemed to have recovered, too, by the time scientists ventured into the exclusion zone on foot in the mid-1990s.

A detailed study published in 2015 backed up these observations. Jim Smith at the University of Portsmouth, UK, and his colleagues in Belarus, the UK, Russia and Germany all concluded that Chernobyl's wildlife is in great shape.

The radioactive fallout was certainly severe enough to have an impact on individual humans and justify the evacuation. It even had devastating effects on the health of many individual animals. But crucially, it wasn't harmful enough to have a lasting effect on the health of wildlife at the population level.

If anything, those populations may have received a significant boost as a consequence of the Chernobyl accident. Wherever there is human activity, there is pressure on wildlife. The forced evacuation of so many people from the Chernobyl exclusion zone gave local plants and animals an opportunity to prosper. By 2016, there were even calls to turn the exclusion zone into a wildlife reserve.

However, some scientists think it's very wrong to conclude that wildlife around Chernobyl is thriving. Two in particular – Anders Møller at the University of Paris-Sud, France, and Timothy Mousseau at the University of South Carolina in Columbia – have also been working in the exclusion zone for many years. They think that the animal populations there are struggling.

Møller and Mousseau have counted fewer large animal tracks inside the exclusion zone than outside, suggesting there are fewer big mammals around Chernobyl. Insects and spiders are suffering too, they say.

Most worryingly of all, Møller and Mousseau report that insects are struggling, even in parts of the exclusion zone where radiation exposure is very low – well below what most scientists would consider an acceptably safe level. Some researchers have reported similar findings in the region around Japan's Fukushima Daiichi nuclear power plant, which went into meltdown in March 2011 as a consequence of the devastating tsunami triggered by the Tōhoku earthquake.

These two viewpoints couldn't be more extreme. Either nature is far more resilient to nuclear disasters than we

hoped, or it is far more vulnerable than we feared.

It's not entirely clear how scientists studying the same animals in the same area could have reached such different conclusions about their well-being – although some researchers suspect they know where the problem lies. They say that these studies should ideally be carried out by radioecologists – scientists who have been trained how to study the effects of radiation on the environment. Many are actually carried out by general ecologists who can unwittingly misinterpret radiation data.

Larger, coordinated studies involving many scientists – including radioecologists – should help establish which of these very different narratives is correct. Until a consensus view emerges, though, it's very difficult to properly assess the lasting legacy of the Chernobyl disaster.

(Opposite) The city of Pripyat was abandoned following the Chernobyl nuclear accident.

(Above right) Some scientists say large mammals are now thriving in the Chernobyl exclusion zone.

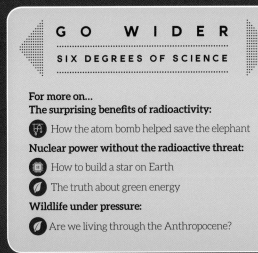

GO WIDER

SIX DEGREES OF SCIENCE

For more on…
The surprising benefits of radioactivity:

How the atom bomb helped save the elephant

Nuclear power without the radioactive threat:

How to build a star on Earth

The truth about green energy

Wildlife under pressure:

Are we living through the Anthropocene?

THE KILLERS LURKING IN EARTH'S ICE

Giants don't come much smaller than this. In 2014, biologists discovered what was at the time the world's largest virus – a monster measuring one-and-a-half thousandths of a millimeter in length.

Our planet is losing more of its ice and permafrost every year.

It wasn't just the size of Pithovirus that was astonishing, it was where it was found. It was resurrected from 30,000-year-old permafrost. And although this particular virus is harmless to humans, others in the frozen soil might not be – which raises a worrying question.

What might be unleashed as Earth's natural freezer begins to warm up?

Global temperatures are rising year on year, largely as a consequence of human-driven climate change. Those warmer conditions pose a real threat to the continued existence of icy deposits near the poles and in high-altitude areas.

As the ice and permafrost begins to melt, all sorts of

ancient frozen treasures are coming to light.

Probably the most famous example is the astonishing 5,000-year-old "iceman," dubbed Ötzi, discovered in the Alps in 1991. His body is so exceptionally well preserved – tattoos, rotting teeth, intestinal parasites and all – that it has provided archaeologists with an unprecedented window into the life of a European herdsman during the Copper Age. It has provided an insight into the death of a herdsman, too: forensic studies suggest Ötzi was murdered before he entered his natural icy tomb.

Extinct Ice Age mammals are also emerging from the frozen ground. A 40,000-year-old mammoth discovered in Siberia in 2013 is one of the most complete and well preserved ever found. The carcass reportedly oozed liquid blood as it was examined, while the meat remained so fresh that one scientist claimed to have taken a bite.

Some geneticists think the specimen might preserve enough genetic material to move forward with controversial efforts to resurrect the iconic beasts. By 2015, there were reports that Harvard geneticist George Church had spliced mammoth genes for long hair and relatively

(Above left) "Buttercup," a 40,000-year-old mammoth discovered in Siberia in 2013.

(Above) Ötzi the iceman is still revealing secrets about life in prehistoric Europe.

small ears into the DNA of cells from Asian elephant skin.

That's still a very long way from actually bringing back the mammoths, but extinction might ultimately be reversible.

Seeds that were apparently gathered and buried by foraging squirrels 30,000 years ago have also come out of Siberian permafrost. One 2012 study, led by David Gilichinsky at the Russian Academy of Sciences in Pushchino, reported that some of the seeds germinated and grew – although some scientists are skeptical that the seeds really are as old as is claimed.

However, there is no such skepticism over Pithovirus. This particular icy resident popped out of a sample of ancient frozen soil from Siberia. Not only had the virus survived intact for that length of time, it could be coaxed back into an active state. Large viruses often attack amoebas – Jean-Michel Claverie at Aix-Marseille University in France and his colleagues found that their Pithovirus would still attack amoebas and reproduce.

Some members of Claverie's team told the media that viruses with the potential to harm humans might well survive in the permafrost, too.

There may be something even more deadly than viruses trapped in the frozen earth, though. A huge amount of carbon is locked away in permafrost. As the ground thaws, the carbon can seep out and enter the atmosphere as carbon dioxide and methane – both powerful greenhouse gases. The gases will warm our planet a little more, which will melt more permafrost and release more greenhouse gases – and so on.

Some fear the consequence of this feedback loop. They worry that defrosting Earth's natural freezer will accentuate the already challenging problems of climate change.

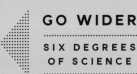

GO WIDER

SIX DEGREES
OF SCIENCE

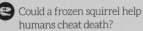

For more on…
Organisms that can survive freezing:
Could a frozen squirrel help humans cheat death?

Reversing extinction:
Dinosaur resurrection: meet the chickenosaurs

The battle against climate change:
Are we living through the Anthropocene?

When life kills itself

The microbes that eat (and poo) electricity

THE TRUTH ABOUT GREEN ENERGY

Very slowly, the world is weaning itself off its fossil fuel addiction. Wind and solar farms are replacing coal-fired power plants, and electric cars are beginning to push gas vehicles off the roads. Our planet will be a greener, more environmentally friendly place as a result.

But perhaps this is too optimistic. Some renewable power technologies are not quite as green as they appear.

The idea that renewable power might not be particularly green will come as no surprise to some environmentalists. They have long campaigned against nuclear power because of the radioactive waste it generates. However, many might be more surprised to learn that non-nuclear green technologies have environmental problems of their own.

There's one key reason behind these environmental

problems. Green technology has a hunger for obscure and hard-to-source elements.

Few people will have heard of neodymium, for instance – although many of us hear *with* it. Even in small quantities, neodymium is powerfully magnetic, making it an ideal material to fashion into the tiny electromagnets that help in-the-ear audio headphones produce sound. This magnetism also makes neodymium very useful in wind turbines and electric cars, both of which rely on strong magnets to function.

Sixteen of the 17 rare earth elements. ·················

(Opposite) Wind farms generate clean energy, but manufacturing them can be environmentally costly.

Energy-efficient LED light bulbs, meanwhile, generally require special chemical compounds containing elements like yttrium and terbium to make sure that the light they emit has a warm and pleasant hue – and not a cold and clinical one.

All of these unfamiliar chemical elements belong to a special class called the "rare earths." It's a confusing name – rare earths are not particularly rare. But they seldom occur in convenient high-concentration deposits: mining companies have to extract huge quantities of material to produce small amounts of the rare earth elements. Worse, that material must be treated with a cocktail of chemicals, including some toxic ones, to pull the rare earths out.

It's costly work, both from an economic and an environmental perspective. In the recent past, it's really only been companies based in China and elsewhere in southeast Asia that have been prepared to mine for rare earths – and some reports suggest local environments in the region have suffered as a consequence. With so few companies controlling the world market, some governments have expressed concerns that supplies of critical rare earths might begin to dry up in a few years.

Even if other countries decide to begin mining rare earths, too, it's doubtful they could do so in a truly environmentally benign way and still compete economically with mines in southeast Asia.

Something needs to change.

Perhaps, eventually, miners will be able to tap into new sources of these elements in space rather than on Earth. Not only would this help to deal with supply issues, it could also effectively transfer the environmental impact of extraction off the planet. But the nascent space mining industry isn't yet ready to begin hunting down asteroids and stripping them of their natural resources.

In the shorter term, the renewable power industry – and technology companies in general – will probably have to become a lot greener by improving their record on rare earth recycling. A UN report in 2011 found that neodymium is among a surprisingly large number of elements with a recycle rate below one percent. As recently as 2013, the organization was still calling for a "rethink" on recycling.

At least one study suggests recycling might be worth it from an environmental perspective. In 2014, a team led by Benjamin Sprecher at the Materials Innovation Institute in Delft, the Netherlands, calculated that some forms of rare earth recycling – in this case extracting neodymium from computer hard drives – could reduce the environmental impact of sourcing the metals by a factor of ten.

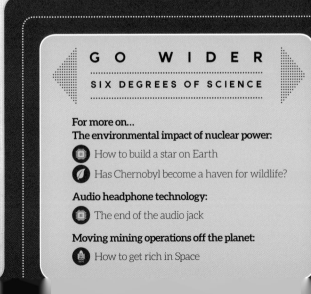

GO WIDER

SIX DEGREES OF SCIENCE

For more on...
The environmental impact of nuclear power:

How to build a star on Earth

Has Chernobyl become a haven for wildlife?

Audio headphone technology:

The end of the audio jack

Moving mining operations off the planet:

How to get rich in Space

CAN SUPER-CORAL SAVE OUR SEAS?

A formerly colorful world is now rendered in stark monochrome. Unusually warm ocean temperatures between 2014 and 2016 triggered the longest and largest coral bleaching event since records began, turning vibrantly colored corals to deathly white.

Almost 90 percent of the coral in some areas felt the effects, severely limiting their ability for healthy growth. It's just the latest worrying evidence that coral can't cope with life in our rapidly warming world.

Help might be at hand, though. A plan to artificially speed up the evolutionary process could result in a new breed of super-corals, far more resilient to the effects of climate change.

Human-induced climate change has not been kind on coral. Many coral species require a very narrow range of conditions for healthy growth: the right amount of sunlight, the right temperature and the right water chemistry.

Climate change has rocked their world in two ways. Most obviously, ocean temperatures have begun to rise at a rapid rate. But there's another factor: the carbon dioxide being pumped into the atmosphere forms a weak acid when it dissolves in seawater. This changes the ocean's pH, putting all kinds of marine species under stress in the process.

One 2016 study led by Nancy Muehllehner at the University of Miami, Florida, found evidence that some of the coral around the Florida Keys might actually have begun to dissolve away because of this "ocean acidification."

However, there's still a huge amount of diversity among corals, and some of them seem to be genetically tougher in the face of climate change than others. That observation gave two biologists – Ruth Gates at the Hawaii Institute of Marine Biology in Kaneohe and Madeleine van Oppen at the Australian Institute of Marine Science in Townsville – an idea. What if they could identify the hardiest individuals and selectively breed them to produce new lines of corals that can cope with climate change?

The idea is not as far-fetched as it might sound: after all, humans have been selectively breeding plants and animals ever since they began farming thousands of years ago. It's only because of such selective breeding that most of us have enough food to eat.

Perhaps in recognition of the tantalizing promise of the

Unusually warm water can trigger coral bleaching events.

Spawn from hardy corals could be used in selective breeding programs.

plan, Gates and van Oppen received a $4 million (£3.14 million) grant in 2015 to explore the idea further.

Part of the plan is to swim over the reefs affected by the recent bleaching in search of the few corals that, almost miraculously, managed to stay healthy. Selectively breeding these corals should help make the hardier corals even tougher.

Gates and van Oppen are also exploring other ways to toughen up corals. One approach is to expose hardy corals to even more challenging conditions to try to accelerate their response to climate change.

For instance, in one 2015 study, Gates and another colleague – Hollie Putnam – collected pregnant adult coral and moved them into the lab. They kept some of the coral in normal water conditions and some of them in the warmer, more acidic water that scientists predict will become the norm in future.

Then, Gates and Putnam collected the youngsters released by the two groups of coral and explored their environmental tolerances. It turned out that the youngsters born of stressed corals could cope better with the warmer, more acidic water than the youngsters birthed by unstressed corals.

An Intergovernmental Panel on Climate Change report in 2014 concluded that coral reefs were Earth's "most vulnerable marine ecosystem, with little scope for adaptation." The work being spearheaded by Gates and van Oppen suggests that there might be room for a glimmer of optimism. Coral might be far more adaptable than most marine biologists assume.

GO WIDER

SIX DEGREES OF SCIENCE

For more on...
Combating climate change:

 Are we living through the Anthropocene?

 Saving the planet one eruption at a time

WHEN LIFE KILLS ITSELF

Evolutionary history is written by the victors. Between two and three billion years ago, life on Earth made a significant breakthrough. One type of bacteria evolved a revolutionary new form of photosynthesis that kicked out oxygen gas as a waste product. With time, the oxygen level in Earth's atmosphere rose, paving the way for the evolution of complex life forms including animals and, ultimately, human beings.

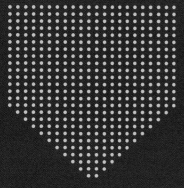

Living creatures could help wipe out species and make Earth uninhabitable.

Cyanobacteria pumped "toxic" oxygen into Earth's early atmosphere.

But all those billions of years ago, the evolution of oxygen-generating bacteria was very bad news for the other microbes alive at the time. All of these were "anaerobic" organisms. As far as they were concerned, oxygen was toxic – and many must have died in what may have been one of the biggest extinctions of life in the history of our planet.

The event – sometimes called the "Oxygen Catastrophe" – is held up as a classic example of life's uncanny ability to turn into its own worst enemy.

For the second half of the twentieth century, many people believed that organic life interacts with its inorganic environment in a beneficial and self-perpetuating way: life ensures it own continued survival. It's an idea called the Gaia hypothesis.

The Gaia hypothesis offers an explanation for why Earth and its atmosphere are dynamic, vibrant and alive. Mars, in contrast, has a thin and stagnant atmosphere and a dry, dusty surface, probably because it carries little or no life.

Many scientists, though, don't believe in Gaia. In 2009, one biologist – Peter Ward at the University of Washington in Seattle – came up with a counter theory.

He says the history of life on Earth might seem "Gaian" from a distance, but look closer and it is littered with chapters – like the Oxygen Catastrophe – in which life itself has managed to bring about a mass extinction of life.

There have been five big (and many more small) mass extinctions over the last 500 million years. Most seem to coincide with large-scale volcanic activity. This often belches vast quantities of carbon dioxide into the environment and triggers global warming.

Such warming is, in itself, bad news for many species. But Ward says the real devastation often comes a little later. As the Earth's poles become similar in temperature to the tropics, ocean currents weaken and ocean water stagnates. In those stagnant conditions, there is no way to replenish oxygen supplies deep in the ocean. Most life forms suffocate, except for the few that can breathe without oxygen. These include bacteria that pump out poisonous hydrogen sulfide. As oxygen levels fall –

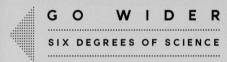
Trilobites, a group of marine arthropods, may have been poisoned to the point of extinction.

and hydrogen sulfide levels rise – more and more organisms die off.

This is the opposite of what the Gaia hypothesis would predict. On Gaian principles, biological ecosystems should respond to problems like global warming by finding ways to pull climate conditions back into a state that is beneficial to most life forms. What really seems to happen is that the deteriorating conditions encourage the growth of some organisms that have the power to make conditions even tougher for the rest of the ecosystem.

Ward calls this more brutal view of life the "Medea hypothesis." While Gaia is the ancient Greek goddess of the Earth, Medea is a character from Greek mythology who killed her own children.

In the foreseeable future, Medea could raise her head again.

Today's global climate is warming at an alarming rate, with polar temperatures rising particularly fast. Ocean currents could soon begin to weaken. Ocean stagnation would follow, and that would set up perfect conditions for the bacteria that pump poisonous hydrogen sulfide into the atmosphere. It's conceivable that human civilization will one day be brought crashing down through the action of humble bacteria.

GO WIDER

SIX DEGREES OF SCIENCE

For more on...
Extinction:

📖 The end of everything

The warming climate:

🌓 Are we living through the Anthropocene?

🌓 The killers lurking in Earth's ice

CHAPTER FIVE

NATURAL WORLD

THE MICROBES THAT EAT (AND POO) ELECTRICITY

There is a power grid beneath the seafloor. It's hooked up in a way that keeps methane – a potent greenhouse gas – locked within the seabed where it can't enter the atmosphere and accelerate global warming.

It sounds like the end result of a hugely ambitious engineering effort to reduce the effects of climate change. But the seafloor electricity grid wasn't built by humans. It has grown organically, between microbes that eat and excrete electricity.

Electricity powers the modern world – but for some microbes, it's just lunch.

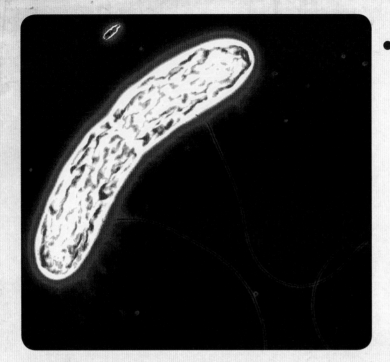

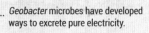
Geobacter microbes have developed ways to excrete pure electricity.

coax these bacteria out of the mud. They congregate around the electrode, sucking up the electrons flowing out of the metal and using them to produce energy inside their single-celled bodies.

They "eat" pure electricity.

These findings are strange enough, but there is more. Yuri Gorby at the Rensselaer Polytechnic Institute in New York and other biologists say that there are some electricity-eating and electricity-excreting bacteria with another talent.

If there isn't an obvious source of electrons in the local vicinity – or an obvious place to dump electrons – these microbes go looking for one. They begin growing thin hairs that stretch away from the cell. Tests show that the hairs conduct electricity: they may act as organic electrical wires to help the microbes hook up to electron sources or dumps.

By 2010, the biologists were uncovering evidence that bacteria don't just hook up to inanimate objects. Sometimes they connect to one another. It's possible – although difficult to definitively prove – that vast colonies of bacteria link together to form organic power grids, with electricity flowing from microbe to microbe.

This is essentially what Antje Boetius at the University of Bremen in Germany and her colleagues believe they found in 2015.

It has long been a mystery exactly how microbes living in the seabed can so effectively prevent methane in the sediment from leaching out and escaping into the atmosphere. Biologists know that some of the microbes living in the seabed can break down the methane, but it's a

The idea of life that survives on nothing but electricity sounds foreign, even alien. In truth, it is not.

Deep down, all life forms operate in this way – even humans. We eat sugar-rich food and breathe in oxygen. Inside our cells, electrical particles – electrons – are stripped from that sugar, and then flow through molecular machinery to help produce the special energy-storing molecules that power our bodies. It is oxygen that allows that electrical current to flow, because it readily takes up electrons.

In the 1980s, biologists – including Kenneth Nealson at the University of Southern California in Los Angeles – began finding bacteria that do things a little differently. These microbes, living in the sediment at the bottom of lakes and oceans, don't "breathe in" oxygen – or any other gas. To encourage the flow of electricity through their cellular machinery, these bacteria simply dump the electrons out of their single cell and onto iron in their environment.

They "excrete" pure electricity.

Within the last few years, scientists have found other bacteria that do something equally strange. Instead of taking in food and stripping it of its electrons, these microbes simply take in electrons in their raw form. Sticking an electrode into the sediment on a lakebed can

process that generates electrons. For methane breakdown to occur as efficiently as it does, the microbes must have found a very effective dumping ground for those electrons.

Boetius and her colleagues think the microbes have hooked up with electron-eating bacteria that also live in the seabed. Electricity flows from one group of microbes to the other.

This power grid is extraordinary for another reason. The microbes that break down the methane are not bacteria. They belong to a fundamentally different biological domain called the archaea. Bacteria and archaea last shared a common ancestor billions of years

Climate change would be more severe without electricity-eating microbes.

ago. Evolutionarily speaking, they are far more distinct from one another than humans are from oak trees, for instance. But beneath the seafloor, these two types of microbe have learned to work together, giving and receiving electricity.

It's just as well for us that they get on so well. Without the organic power grid they form, global warming would be a far more acute problem.

GENE EDITING JUST GOT SERIOUS

Science rarely gets more contentious than this. In September 2016, there were reports that scientists in Sweden had begun modifying the DNA inside healthy human embryos. The research promises to improve the scientists' understanding of the earliest stages of human life. It may even lead to new ways of preventing infertility issues and miscarriages.

However, for many people, the price to pay for those potential breakthroughs is too high. Editing the genes of human embryos is unnatural and should never be allowed.

This is a valid point. Humans and other animals can't naturally edit their genomes to switch genes on or off. But other organisms can. Gene editing of exactly the kind that was carried out on the human embryos in 2016 has probably been going on naturally for billions of years.

Geneticists have been honing their gene-editing skills for years. The techniques they have developed to switch genes on or off inside living cells are effective – but they are expensive.

A low-cost version of the technology arrived in 2013. It's called CRISPR (pronounced "crisper"), and it has taken the genetics world by storm. Quite an achievement, given that CRISPR is actually a piece of molecular machinery that evolved perfectly naturally in humble microbes.

Bacteria may have been editing their own genomes for billions of years.

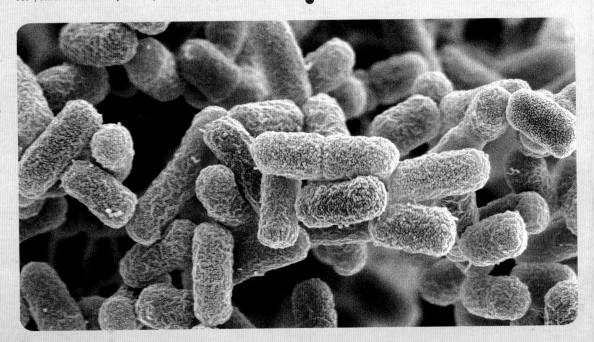

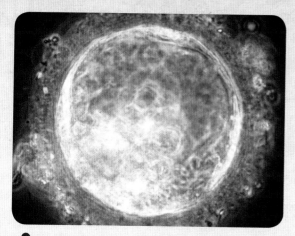

Geneticists are now modifying healthy human embryos.

The CRISPR story begins in the final years of the 1980s. Japanese biologists studying a bacterial genome noticed that there were strange repetitive sequences in the DNA. These "repeater" sequences were separated from each other by short and apparently random chunks of genetic code, which were dubbed "spacers."

By the mid-2000s, scientists had found these spacer and repeater sequences in about one-third of the species of bacteria they examined. In the other great domain of microbial life, a group of single-celled organisms called archaea, they are even more common.

The scientists had also figured out what they are for. They are evidence of a strange and ancient immune system wired into most microbes' DNA.

Microbes are under near-constant attack from viruses. Some estimates suggest viruses kill half the bacteria on the planet every two days – leading some researchers to explore whether viruses can be used in medicine to replace society's increasingly ineffective antibiotics.

But the microbes do not simply accept their fate. Many produce a special type of protein that sniffs out the attacking virus, locks onto its DNA and then chops out a chunk. This snippet of viral DNA is then incorporated into the microbe's own DNA sequence, forming one of the "spacer" sequences. This seems to help the microbe recognize the virus if and when it attacks again, allowing the microbe to mount a speedy defense.

Scientists quickly recognized that the microbes have essentially evolved a DNA cut-and-paste tool, which is exactly the sort of thing that geneticists have spent years developing. But the microbial version is both elegantly simple and cheap to modify into a tool that can cut any given piece of DNA at a precise point in the genetic sequence – which is why CRISPR is so popular.

There's one more twist to the CRISPR story. Biologists who study evolution are also interested in CRISPR, for a completely different reason.

CRISPR, they realized, offers bacteria a way to hardwire life experiences into their very DNA, creating a permanent record of the viruses they have encountered. Because a microbe's offspring inherit that DNA, those life experiences are passed down the generations.

Arguably, this makes the CRISPR immune system superior to the one that operates inside the human body. We have to experience a viral infection ourselves before we can gain immunity – we don't inherit immunity from our parents.

For decades, biologists assumed that life experiences can't possibly be inherited: the CRISPR system in bacteria shows they can be in some cases. It is changing the way biologists think about evolution.

It's not the first time microbes have astonished evolutionary biologists. Scientists once thought evolution was all about "vertical" inheritance – genes passed down from an adult to its offspring. But microbes have also developed a sophisticated system of "horizontal" evolution: they can swap useful genes with any microbe they meet. This process helps explain why resistance to antibiotics spreads so rapidly through bacterial populations.

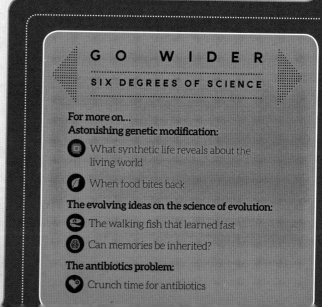

GO WIDER

SIX DEGREES OF SCIENCE

For more on...
Astonishing genetic modification:

- What synthetic life reveals about the living world

- When food bites back

The evolving ideas on the science of evolution:

- The walking fish that learned fast

- Can memories be inherited?

The antibiotics problem:

- Crunch time for antibiotics

WHY CANCER IS LIKE A SELFISH ANIMAL

Hundreds of millions of years ago, a revolution was unfolding. Cells were learning to cooperate. They were evolving ways to stick together – literally – and to put selfish individualism to one side in favor of communal living. Earth was witnessing the birth of animal life – and, just possibly, the birth of cancer.

Earth is about 4.54 billion years old. For most of that time, it has been inhabited: in 2015, Elizabeth Bell and Mark Harrison at the University of California in Los Angeles and their colleagues found tantalizing signs of organic life in rock samples that are an astonishing 4.1 billion years old.

However, it was an exceptionally long time before life in its most familiar form began to appear. For billions of years, the only living things were tiny, single-celled microbes. Animals, which are large and complex creatures, began to appear only about 650 million years ago.

It's still a mystery why it took animals so long to evolve. Many geologists link their late appearance to changes in Earth's atmosphere – in particular, to a rise in the level of oxygen gas, which animals breathe.

Some biologists disagree. They think animals didn't evolve until 650 million years ago, simply because they are such complicated organisms: it just took evolution a very long time to "work out" how to evolve animals.

There are all sorts of reasons why animal life might have been difficult to evolve. One factor that perhaps shouldn't be overlooked is an obvious one: animals are composed of many cells.

Until animals appeared, life had largely been content to work on the level of single cells, all working independently to maximize their own chances of survival. Animals broke that ancient rule – an animal body can contain trillions of cells. Those cells have had to learn to put aside a natural instinct to go it alone and instead act in the interest of the animal itself.

That willingness to cooperate has been pushed to

Cancer cells are good at multiplying, even though this behavior kills their host.

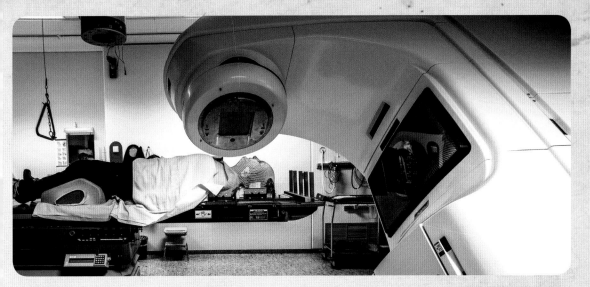

 New cancer therapies could stem from studies into the evolution of cancer.

an extreme: some of our cells will deliberately kill themselves if they pose a risk to our bodies. It's the ultimate selfless act.

A cancer tumor is effectively an animal gone wrong. Its cells cooperate with each other, but they are selfish, too. They stubbornly refuse to kill themselves, even when it becomes obvious that they threaten the health of the body as a whole, which is what makes a cancer diagnosis so worrying.

In 2011, two researchers – Charles Lineweaver at the Australian National University in Canberra and Paul Davies at Arizona State University in Tempe – came up with a new way of thinking about cancer. They suggested malignant tumors are what happens when hundreds of millions of years of animal evolution are reversed. Cancer tumors, they say, behave like the very first animals did.

According to this idea, the first animals were still "learning" how to control the multitude of cells within their bodies. They had worked out how to prevent their individual cells growing and replicating uncontrollably, but they hadn't completely crushed those cells' deeply ingrained selfishness. The first animals were unstable, and their bodies were full of selfish cells. They were basically cancer tumors.

It didn't take long for animals to evolve beyond this stage and develop the genes that fully tamed their cells' selfish instincts. But if those genes mutate in even one human cell, it begins to behave like a cell inside one of those primitive animals. It develops into a cancerous tumor.

It's fair to say that there are plenty of scientists who take issue with this idea. But there is nevertheless a growing interest in seeing cancer through the prism of animal evolution. As scientists learn more about the ways that animals evolve, the hope is that they might open up new fronts in their battle against one of the most deadly of animal diseases.

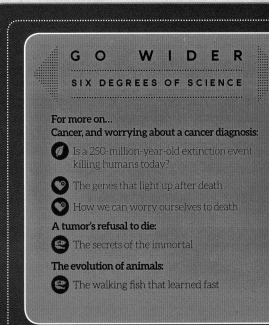

GO WIDER

SIX DEGREES OF SCIENCE

For more on...
Cancer, and worrying about a cancer diagnosis:

Is a 250-million-year-old extinction event killing humans today?

The genes that light up after death

How we can worry ourselves to death

A tumor's refusal to die:

The secrets of the immortal

The evolution of animals:

The walking fish that learned fast

WHAT LIVES IN THE FOURTH DOMAIN?

It's a crowded planet. A 2016 estimate suggests Earth may be home to upwards of a nonillion (one followed by 30 zeros) individual organisms, living in a trillion or so different species – and that's just the microbes. As far as biologists know, all of the great diversity of life falls into just three great categories, or domains.

The question some are now asking is whether they are on the verge of identifying a brand new fourth domain of life.

Ever since Darwin, scientists have suspected that all life on Earth stemmed from a single common ancestor – a microbe nicknamed LUCA that lived more than 3.5 billion years ago. By about two billion years ago, LUCA had given rise to three distinct kinds of organism: the bacteria, another group of single-celled microbes called the archaea, and a third group – the eukaryotes – that later evolved into all plants, animals and fungi.

Every single organism that biologists have ever found belongs to one of these three vast groups. Most scientists think that's because there are only three domains of life

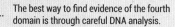
The best way to find evidence of the fourth domain is through careful DNA analysis.

on Earth. However, some suspect a fourth domain still lurks somewhere out there, just waiting to be found.

They have their reasons for thinking this way. Although scientists have so far found only bacteria, archaea and eukaryotes in their surveys of life on Earth, the number of species those surveys cover is relatively small – about a million. If the estimates are correct and there really *are* a trillion species in total, that leaves an awful lot of undescribed species out there. Not all of them might fall into one of the established three domains.

But finding evidence of a fourth domain of life is very challenging. If it exists, it almost certainly consists entirely of microbes, rather than species big enough to see with the naked eye. This presents biologists with a problem because – surprisingly – almost all species of microbe refuse to grow in the lab, which makes it virtually impossible to study them using traditional scientific methods.

There is now an alternative way to look for the fourth domain, though. Biologists can take a sample of material from the environment – a bucket of seawater perhaps, or a shovel-load of soil – and put it through state-of-the-art lab equipment that sequences the DNA of any organisms the sample contains. If some of that DNA looks drastically different from anything scientists have ever seen before, it might suggest that it came from microbes belonging to the mysterious fourth domain.

In 2011, biologists including Jonathan Eisen at the University of California in Davis and Craig Venter of the J. Craig Venter Institute in Rockville, Maryland, found

 Could fourth domain microbes be lurking inside the human gut?

some DNA sequences that fit the bill. Their team sequenced DNA from ocean water samples and discovered two gene groups that were quite unlike any seen before. It's possible that both gene groups came from a fourth-domain microbe that lives somewhere out in the ocean.

Unfortunately, sifting through the world's seawater to actually identify this tiny microbe would be considerably harder than finding a needle in a haystack.

In 2015, though, scientists reported possible signs of the fourth domain in a much more confined space: the human body.

The human gut is home to a massive array of reasonably well-studied microbes – a community called the "gut microbiome." Eric Bapteste and his colleagues at the Pierre and Marie Curie University in Paris, France, sequenced all of the microbial DNA in gut samples, and searched

through it specifically on the lookout for genetic sequences that were drastically different from any seen before.

They found them: thousands of DNA sequences that were nothing like any ever encountered before.

Results like this are exciting and intriguing. However, biologists won't know for sure that such strange DNA really does come from fourth-domain microbes, unless and until they can find and study the microbes to which the DNA belongs.

The good news is that they might be able to use the strange DNA to fish them out. And knowing that they live in a relatively confined environment – the human gut – should make the task a little easier.

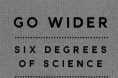 GO WIDER

SIX DEGREES OF SCIENCE

For more on...
Life's fundamental divisions:

The microbes that eat (and poo) electricity

The discoveries being made inside the human body:

 The life-saving power of excrement

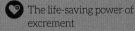 The secrets hidden inside your body

THE DWARF DINOSAURS OF TRANSYLVANIA

In January 2016, scientists unveiled a brand new dinosaur species at the American Museum of Natural History in New York. Even by dinosaur standards, the then-unnamed species was astonishing: at almost 40 meters (131 feet) long and 20 meters (66 feet) tall, it is the largest land animal ever found.

Seeing such a formidable animal alive must have been a spectacular sight. However, unless schemes to resurrect dinosaurs somehow succeed, we must rely on our imaginations – and computer-generated graphics – to put flesh on such extraordinary bones.

Some dinosaur fossils from the Transylvania region of Romania might offer clues to explain why giant dinosaurs no longer exist. In this corner of eastern Europe, scientists have discovered fossils every bit as remarkable as those on display in New York. They belong to dwarf dinosaurs.

The way scientists think about dinosaurs has changed dramatically in just a few decades. What were once seen as scaly, sluggish monsters are now reconstructed as brightly colored and often swift-footed creatures.

One thing hasn't changed, though: dinosaurs haven't lost their reputation for size.

Large body size made sense both for herbivorous and carnivorous dinosaurs. It offered herbivores protection from predators, and it gave those predators the powerful jaws they needed to have any chance of bringing down a large herbivore.

But not all dinosaurs were giants. Some were knee-high to famous beasts such as *Tyrannosaurus* and *Diplodocus*. Since 2006, fossils of these dwarf dinosaurs have been found in Germany, the UK and even in Saudi Arabia.

The most interesting of all dwarf dinosaurs are those from Transylvania. Here, right at the end of the dinosaurs' long reign, there was an entire community of dwarfs. A tiny, long-necked, long-tailed herbivore called *Magyarosaurus* formed herds, as did a relatively small,

Was this the largest dinosaur that ever walked the Earth?

duck-billed dinosaur named *Telmatosaurus*. Predators lived in the area, too, including the tiny, two-legged *Balaur*, named by paleontologists in 2010.

All of these Transylvanian dinosaurs were trapped on a large island, because the global sea level was so high at the end of the dinosaur era that most of continental Europe was submerged. It was almost certainly island life that made the dinosaurs small.

There's something about an island's confined space and relative lack of food that encourages some animals

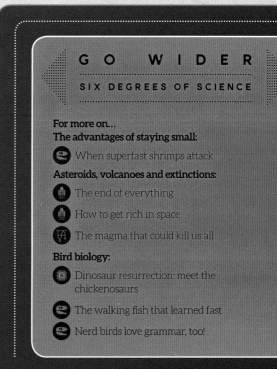

When disaster struck, even dwarf dinosaurs were too large to survive.

to evolve unusual body sizes. It happened to elephants and hippos that lived on Mediterranean islands tens of thousands of years ago. It even happened to a now extinct relative of our human species: on the Indonesian island of Flores, ancient humans shrank and evolved into 1-meter- (3-foot-) tall "hobbits."

But what is perhaps most interesting about the dwarf dinosaurs of Transylvania is that they never really became that small. Many of these "dwarf" dinosaurs were still about as big as a cow – which, by modern standards, makes them rather large. It seems that, for reasons scientists still can't really explain, dinosaurs were just not very good at being small.

Roger Benson and his colleagues at the University of Cambridge, UK, confirmed as much in 2014. They compiled all the information then available on dinosaur body size. They realized that almost no dinosaur species had an estimated weight below one kilogram (2.2 pounds). Today, eight in every 10 species of mammal weighs less than a kilo (2.2 pounds).

There is a glorious exception to this large dinosaur rule. One group did work out how to be truly small. Even tens of millions of years ago, when giants like *Tyrannosaurus* and *Triceratops* were in their prime, the skies were filled with relatively small birds – which scientists now know are dinosaurs.

These contrasting patterns of body size might have been a deciding factor in the famous dinosaur extinction event.

About 66 million years ago, volcanic activity and an asteroid impact helped trigger widespread environmental turmoil that wiped out all the large species on Earth. For some reason, smaller species weren't as badly hit, which

might explain why birds – and our small mammalian ancestors – made it through the disaster.

If other dinosaur groups had shared the birds' ability to evolve into truly tiny forms, living descendants of the giant dinosaur now on display in New York might still walk the Earth. But the dinosaurs of Transylvania tell us that there seems to have been some constraint that meant even dwarf dinosaurs were just too big to survive.

G O   W I D E R

SIX DEGREES OF SCIENCE

For more on...
The advantages of staying small:

When superfast shrimps attack

Asteroids, volcanoes and extinctions:

The end of everything

How to get rich in space

The magma that could kill us all

Bird biology:

Dinosaur resurrection: meet the chickenosaurs

The walking fish that learned fast

Nerd birds love grammar, too!

THE WALKING FISH THAT LEARNED FAST

How's this for adaptability: a type of fish called *Polypterus* has learned to "walk" on land by lengthening and strengthening some bones inside its skeleton. In doing so, it has mirrored one of the most iconic events in the last half a billion years of evolutionary history – the moment our distant ancestors left the water and crawled onto land.

That sort of evolutionary event should play out gradually over thousands or even millions of years. But *Polypterus* learned to walk in a fraction of that time. In fact, it took just eight months.

Careful biological work has helped show evolution in action, confounding expectations of those who refuse to accept Darwin's ideas. For example, biologist Richard Lenski of Michigan State University in East Lansing has shown that lab populations of bacteria can evolve to eat nutrients that their ancestors couldn't process.

But, critics point out, this is still evolution at the microscale: Lenski's bacteria may have evolved, but they are still bacteria.

In 2014, Emily Standen at the University of Ottowa in Canada and her colleagues published results of a study that begins to answer that criticism. It offers hints of how evolution at a much larger scale might occur.

Standen and her colleagues took young *Polypterus* fish and put them in an aquarium that had been drained of nearly all of its water, forcing the fish to scrabble along the sediment on the aquarium floor in order to move

around. The scientists could perform this sort of "fish-out-of-water" experiment because *Polypterus* is one of a small number of fish that can breathe air: it doesn't suffocate if it is forced to live on land.

In a matter of months, something extraordinary began to happen. The fish learned to walk. They held their paired fins closer to their bodies, allowing them to raise their heads off the ground. The bones at the base of those fins – around the fish's "shoulders" – became larger and stronger. And the "neck" bones became slightly weaker, giving the fish some ability to move their heads independently of their bodies, just like most land animals can.

What makes these findings most exciting is that scientists have seen them before. In 2006, paleontologists unearthed 375-million-year-old fossil remains of an animal – called *Tiktaalik* – that seems to capture the very moment our distant fishy ancestors left water and began

Polypterus began using its fins to "walk" in a matter of months.

Tiktaalik lived right at the moment in prehistory that our ancestors first left the water.

to walk on land. The shoulder and neck bones of *Tiktaalik* show similar adaptations to those seen in the *Polypterus* fish in Standen's lab.

Most scientists think this sort of dramatic evolutionary event – one that turns a swimming fish into a walking animal – happens slowly, over the course of countless generations. The *Polypterus* experiment suggests an alternative. Perhaps animals have the versatility to adapt to dramatic changes in their environment in just a few months or years.

There are several more examples that fit this idea. For instance, birds – which are a form of dinosaur – effectively wind back their evolutionary clock in weeks if they are fitted with long, dinosaur-like prosthetic tails after hatching. They grow up walking in the way that large prehistoric dinosaurs are believed to have done.

Technically, this isn't evolution as scientists recognize it. Evolution is about a permanent change to an animal species through modifications to its genes. The *Polypterus* fish may have learned to walk, but their DNA hasn't changed. If their offspring are put back in a water-filled aquarium, they will grow up swimming, not walking.

However, perhaps the walking fish have taken a first step towards real, genetic evolution. Imagine if Standen's team allowed their walking fish to breed, then raised the young in the same, almost empty, aquarium. These youngsters would probably also learn to walk. And if their offspring, in turn, were raised in the same aquarium, they would learn to walk, too.

Now imagine repeating this process over hundreds of generations. As far as the fish are concerned, walking will have become their natural means of getting about.

Under those circumstances, any fish whose DNA changes in a way that makes walking a little bit easier will have a better chance of survival than its peers. Over time, the fish might pick up so many "walking-friendly" DNA changes that walking becomes hardwired into their genes.

It's almost as if the adaptability displayed by Standen's fish provides a pencil sketch of what's possible – walking – which DNA could gradually ink in to make the trait a permanent feature of the animals' biology.

Biologists are still debating whether this sort of scenario really plays out in nature. This makes the *Polypterus* experiment significant for yet another reason. It is a reminder that scientists may still have much to learn about evolution, and the way it has produced the variety of animals that swim and walk the Earth.

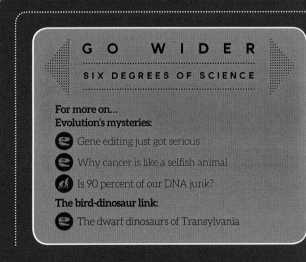

GO WIDER

SIX DEGREES OF SCIENCE

For more on...
Evolution's mysteries:

Gene editing just got serious

Why cancer is like a selfish animal

Is 90 percent of our DNA junk?

The bird-dinosaur link:

The dwarf dinosaurs of Transylvania

THE SECRETS OF THE IMMORTAL

Looks can be deceiving. The Greenland sharks fished from the Atlantic seem relatively unremarkable – but in August 2016, biologists reported that some might be 400 or more years old, making them perhaps the oldest animals with a backbone on the planet.

It isn't obvious yet exactly why the sharks live so long. But the fact that they do hints that they might belong to a very select group of species: the biological immortals.

Despite their special status, biologically immortal organisms are not guaranteed to survive forever. They are as likely as mere mortal species to die through disease or at the hands of a predator. They might even die because of a catastrophic change to the environment that leaves them unable to find the food they need to survive.

A biologically immortal organism will not, however, die of old age. Its cells function as well in its fiftieth or hundredth year of life as they did in the first year.

There are all sorts of factors that might explain how some organisms achieve this biological immortality. The bristlecone pine trees of the western United States, for instance, seem to have unusual stem cells – a special type of cell found in all plants and animals, each of which acts like a biological factory to churn out new tissue. Bristlecone pine stem cells seem to remain vigorous for far longer than the stem cells of most organisms – which might help explain why the trees can live for 4,600 years or more without showing any obvious signs of aging at the cellular level.

Animals can resist the aging process, too – particularly, it seems, those living in the cold waters of the North Atlantic, where the Greenland shark dwells. A shellfish plucked out of the ocean near Iceland in 2006 is one famous example. Scientists count annual growth rings in shellfish shells to work out how old they are, just like tree rings can help age a tree. The Icelandic shellfish – nicknamed Ming – was born just over 500 years ago.

Bristlecone pines are surprisingly youthful at the cellular level.

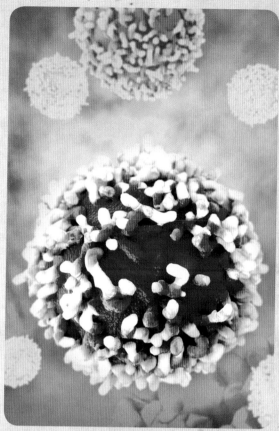

A 2012 study by Daniel Munro and Pierre Blier at the University of Québec in Rimouski, Canada, showed that Ming's cells are unusually resistant to the environmental damage that causes aging. Ming is potentially a biological immortal, or, at least, was. Unfortunately, the scientists didn't realize quite how old and important the shellfish was when they dredged it out of the sea in 2006, and killed it before they began their analysis.

Humans are perhaps uniquely poorly suited to immortality. For instance, some philosophers argue that eternal life would be exceptionally – even perhaps distressingly – boring for a self-aware animal like a human. But at the cellular level, even humans brush with biological immortality.

Most lab cultures of human cells will eventually grow old and die. The DNA inside a human cell decays slightly every time it reproduces itself, and after about 50 cycles of cell division – which usually takes a few days – the DNA decay reaches a lethal threshold.

However, some scientifically famous human cell cultures – called "HeLa" cultures – are now more than 65 years old. These cells generate special molecules that help protect their DNA from the standard decay process. They are biologically immortal.

In this case, though, immortality is a bad thing. HeLa cells are cancer cells that were extracted in the 1950s – without permission – from a patient called Henrietta Lacks.

HeLa cells grow so well that they are now among the most commonly used cells in disease research. Perhaps one day those studies will identify effective ways to make cancer cells mortal, so they are easier to kill. The paradox is that by stripping cancer cells of their immortality, millions of humans would live longer, healthier lives.

(Above left) Ming the Icelandic shellfish was over 500 years old when it was killed.

(Above) Human cancer cells brush with immortality.

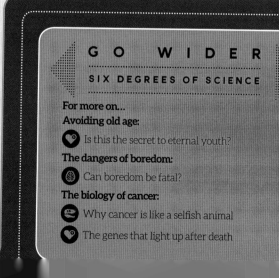

GO WIDER

SIX DEGREES OF SCIENCE

For more on...

Avoiding old age:
Is this the secret to eternal youth?

The dangers of boredom:
Can boredom be fatal?

The biology of cancer:
Why cancer is like a selfish animal

The genes that light up after death

COULD A FROZEN SQUIRREL HELP HUMANS CHEAT DEATH?

Mammals go to great lengths to keep their core body temperature stable. They might pant or sweat to cool down in hot weather, shiver or switch on heat-generating fat deposits to warm up in cold weather. There's a very good reason for doing so: if body temperature strays too far from its normal level, it can prove fatal.

Hibernating mammals break this rule, and none more spectacularly than the Arctic ground squirrel. At most times of year, the rodent's body temperature is about 37 °C (98.6 °F). In winter it drops to -3 °C (26.6 °F).

The squirrel is the only known mammal that lets its body freeze – and it survives the process.

Freezing temperatures are generally very bad news for living organisms. At sub-zero temperatures, water becomes ice, and it typically expands a little as it does so.

The Arctic ground squirrel lets its core body temperature dip below freezing point.

That expansion process can cause delicate biological cells to swell and rupture, just like a water-filled glass bottle might crack if it's left in the freezer by mistake.

The few animals that allow their bodies to freeze when winter comes can do so safely only because they have found ways to prevent this sort of ice damage.

Some insects, for instance, embrace the cold. They deliberately encourage ice to form inside their bodies – but they do so in a carefully controlled way that limits this ice formation to the spaces between cells where it won't damage delicate tissues. If done just right, this can allow the animal to withstand winter temperatures far below zero and live to face the spring.

Several species of ocean fish that live around the poles, meanwhile, add special "antifreeze" molecules to their internal fluids. These molecules readily lock on to tiny ice crystals and arrest their growth so they can't become large enough to cause damage. Using these molecules, the fish can swim around quite happily in seawater that is almost two below zero (seawater remains liquid at such low temperatures, because of the salt it contains).

The Arctic ground squirrel doesn't rely on these tricks. Biologist Brian Barnes at the University of Alaska Fairbanks was the first to record that the ground squirrel's core body temperature drops below zero in winter. He checked for evidence of antifreeze molecules in their ice-cool bodies – but they don't produce any.

Instead, the rodent uses a process called "supercooling" to stay alive. Typically, ice first begins to form by condensing around impurities in a cold liquid – "seeds" that can encourage the growth of large ice crystals. By 2004, Barnes and his colleagues had discovered that Arctic ground squirrels seem to "purify" their blood as winter approaches to remove such seeds, so it can remain liquid at lower temperatures.

The rodents can survive for eight months in a near-permanent frozen state, warming up for just a few hours once every two or three weeks. While frozen, their hearts beat just once per minute. Their brains switch to a strange "stand-by" mode in which many connections between brain cells actually wither away. It's damage that should prove fatal, but research in closely related species of hibernating rodents shows that the connections sprout again as the animals stir from their deep sleep, leaving the squirrels with no lasting effects – testament to the mammalian brain's extraordinary ability to grow and adapt throughout life.

Studying animals that freeze reveals some of the extraordinary consequences of evolution. But these studies also raise a tantalizing question: can scientists replicate these processes in humans?

Space agencies would love to be able to place humans in a state of suspended animation similar to that seen in Arctic ground squirrels – the ability to do so would be a boon for any future deep space missions. Unfortunately, achieving this goal will be enormously challenging: by 2014, Barnes and his colleagues had begun studying how the rodents behave at the cellular level as they prepare for winter. They found that the process is so complex that it involves changing the activity levels of some 500 different genes.

A more realistic goal might be to try to replicate the processes some animals use to protect their tissues from ice damage during freezing. The hugely controversial cryonics industry believes it has done just that.

Its advocates claim that, by pumping medical-grade antifreeze solutions through a recently dead human body, they can protect it from ice damage. Then the body can be put into deep freeze until some future date when medical science will have advanced enough to reanimate the cadaver and restore it to vigorous, youthful health – effectively helping some people cheat death.

At least, that's the hope. The overwhelming scientific consensus is that it is impossible to freeze something as large and complicated as the human body without damaging it in a way that would make it extremely challenging to ever restore it to life. For the moment, whole body freezing is still best left to the animal experts.

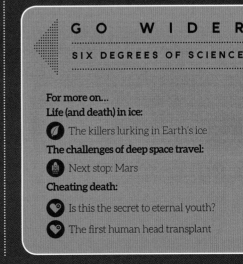

Cryonics companies say they can cool human cadavers without damaging them.

GO WIDER

SIX DEGREES OF SCIENCE

For more on...

Life (and death) in ice:

The killers lurking in Earth's ice

The challenges of deep space travel:

Next stop: Mars

Cheating death:

Is this the secret to eternal youth?

The first human head transplant

WHEN SUPERFAST SHRIMPS ATTACK

The mantis shrimp has an unbelievably quick attack move. Its club-shaped mouth parts accelerate faster than a speeding bullet. They move so quickly that they help generate shockwaves as hot as the surface of the Sun. You might imagine that one of the fastest biological weapons on the planet would be directed towards a rapidly moving target.

If so, you would be wrong. It's often slow-moving snails or completely immobile oysters that are in the firing line.

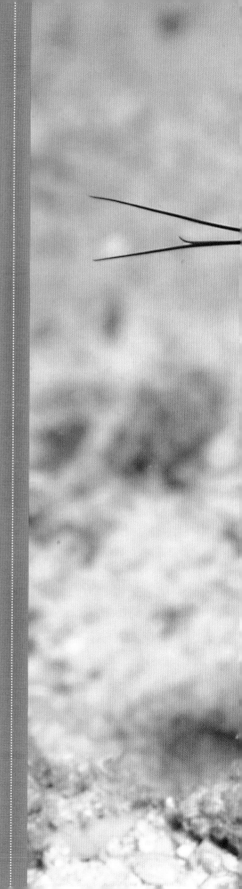

Few animals are as quick on the draw as the mantis shrimp.

Mantis shrimps fall into two broad groups based on the weapons that they use to attack their prey. "Smashers" have big clubs; "spearers" have sharp, arrow-like structures. Both types of shrimp are extraordinarily fast.

They are so fast, in fact, that it's only in the last few years that scientists have had the opportunity to study their attack behavior in any detail. High-speed cameras that can take several thousand frames each second help to slow down the action for detailed analysis.

In 2004, Sheila Patek – now at Duke University in Durham, North Carolina – and her colleagues showed that a smashing mantis shrimp's clubs shoot out at a remarkable 104,000 meters per second squared (341,200 feet per second squared). That's about 7,000 times the peak acceleration rate a cheetah can achieve.

It's physically impossible to accelerate a muscle this quickly. The secret to the shrimp's accelerative speed is a piece of flexible material shaped a little like a potato chip in its tough external skeleton. The shrimp squeezes this material, compressing it. Then, when the shrimp releases the "chip," it instantly springs back into its original shape at an explosive speed. When that power is focused into moving a small structure – the shrimp's mouth parts – the result is an astonishingly fast acceleration.

Similar elastic structures have evolved independently in other species, including ants and jellyfish. Some fungi have them, too: they use them to launch their spores high into the air.

In 2012, Patek and her colleagues put the "spearer" mantis shrimp in front of the high-speed camera for the first time. These mantis shrimp stab fast-moving fish. For obvious reasons, Patek and her colleagues expected spearing mantis shrimp would shoot out their weapons at least as fast as their snail-eating cousins.

However, they don't: spearing mantis shrimp are 100 times slower.

Patek thinks she understands why. Speed is a helpful asset for a mantis shrimp, but so is being quick on the draw. A spearing mantis shrimp needs to be able to prime its weapon and fire it quickly when a fish swims by. It has slightly weaker elastic springs that it can quickly and easily compress – but they don't pack a really fast punch.

The smashing mantis shrimp, on the other hand, can

The cheetah is famous for accelerating quickly.

 Growing large, muscular jaws is one way to develop a powerful attack.

take its time, given that its snail prey can't escape quickly – and oysters can't escape at all.

Smashing shrimp actually evolved from spearing shrimp ancestors about 50 million years ago. Over that time they have adapted to attack much slower prey – so why did they evolve a much faster attack to do so? Patek thinks it's all about packing a powerful punch. Snail shells are one of the most unbreakable of all biological materials. To crack one open, the mantis shrimp needs to deliver a forceful hammer blow.

Two things influence the force of an impact: the mass of an object and the speed of its acceleration. Many predators bulk up to deliver a more powerful attack. The mantis shrimp has focused on boosting acceleration instead, but with a similar effect. The peak force it delivers with its club-like mouthparts is actually about the same as the peak force of a large alligator's bite.

With predatory weapons, as with so many other things in life, size isn't everything.

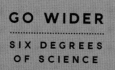

GO WIDER

SIX DEGREES OF SCIENCE

For more on…
Small objects and fast acceleration:
 The tiny spacecraft with big ambitions

Large animals and their powerful jaws:
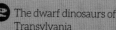 The dwarf dinosaurs of Transylvania

NERD BIRDS LOVE GRAMMAR, TOO!

Pity the strict grammar pedants: they wince at every misplaced apostrophe, every split infinitive, every sentence that begins with a conjunction or ends with a preposition. Some songbirds feel their pain. They are just as intolerant of grammatical blunders in birdsong.

Budgerigars are just one species of bird that listens carefully to what is being sung.

Birdsong is complex – a train of syllables that contains recognizable patterns and repeated phrases. This makes it a lot like human language in terms of its structural organization. And, just like language, birdsong seems to obey some grammatical rules.

If a song breaks those rules, some birds notice.

These findings come from work on the Bengalese finch, a particularly proficient singer. In 2011, biologists Kentaro Abe and Dai Watanabe at Kyoto University in Japan recorded a finch call phrase – essentially an "I am over

here" song – and played it back to other finches until they were familiar with the melody.

Then they snipped up the song into pieces, stuck them back together in a number of different combinations, and played these edited songs to the finches. The birds seemed comfortable with these scrambled sentences – except for one of them. When they heard this particular scrambled song the birds began tweeting away much more furiously than before.

The two researchers speculate that this particular scrambled melody had inadvertently struck a nerve

– broken some cardinal grammatical rule to which the finches reacted. Perhaps most of their scrambled melodies still made some grammatical sense – "Over here I am" or "Am I over here" – but this one particular song just didn't: "Am over I here."

Other studies have explored bird grammar in more detail. A study in 2016 by biologists Michelle Spierings and Carel ten Cate at Leiden University, the Netherlands, found that budgerigars pay close attention to the structure of a bird song "sentence" while a species called the zebra finch simply focuses on the order of particular "words."

The similarities between birdsong and human language might have genetic underpinnings. There are special circuits in the bird brain that are particularly active when a young bird is learning new songs – very similar circuits are switched on during human speech, too.

What isn't clear is why birds and humans share these similarities. They must have developed the traits completely independently, given that humans are mammals and birds are dinosaurs – two groups that haven't shared a common ancestor for hundreds of millions of years.

Curiously, this is not the only thing the two groups may have developed independently. Some birds, like humans, can use tools to solve problems. Both can live in complex social groups. Both even dream when they sleep – which, according to one idea, may be a way for intelligent and social animals to take stock of all the information learned each day.

With luck, exposure to birdsong that is grammatically imperfect doesn't give birds like the Bengalese finch and the budgerigar nightmares.

The Bengalese finch becomes animated when birdsong doesn't make grammatical sense.

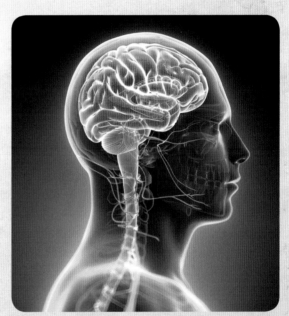

The human brain might share a surprising amount in common with the bird brain.

GO WIDER

SIX DEGREES OF SCIENCE

For more on...
The connection between birds and dinosaurs:

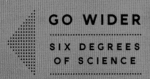 Dinosaur resurrection: meet the chickenosaurs

 The dwarf dinosaurs of Transylvania

The mysteries of sleep:

 Why do we sleep?

CHAPTER SIX

HEALTH & WELL-BEING

CRUNCH TIME FOR ANTIBIOTICS

Where do you turn for a solution to the problem of hospital superbugs? Surprisingly, you could do worse than poring through the pages of a book that was handwritten on animal skin by Anglo-Saxon medics more than 1,000 years ago.

Medicine underwent a revolution in the early decades of the twentieth century. The discovery that Penicillium fungi can kill bacteria ultimately led to the development of penicillin and triggered the birth of the antibiotics industry. Today that industry is in crisis.

Antibiotics are losing their ability to kill bacteria.

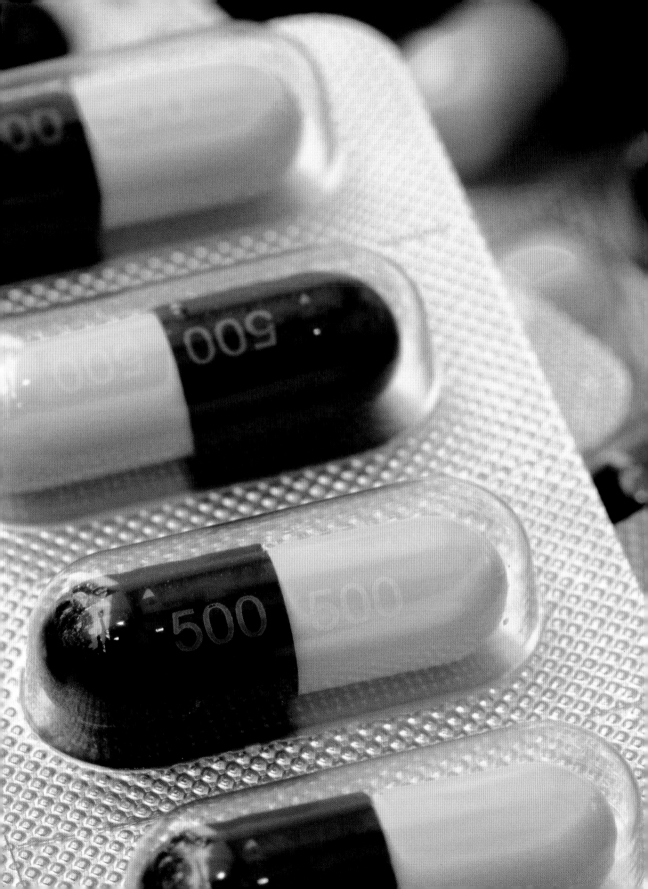

Antibiotics are priceless from a human health perspective, but compared to drugs that tackle cancer or other long-term diseases, the drugs have a relatively low commercial value. Partly as a consequence of this, pharmaceutical companies have gradually shown less interest in developing new antibiotics.

It's an unfortunate trend, because it has allowed bacteria to fight back.

Bacteria reproduce and evolve rapidly. It doesn't take long for some of the microbes in a population to develop resistance to an antibiotic. Those bacteria can quickly spread resistance to other bacteria, because evolution works in strange ways at the microbial level. While human and other animal DNA is passed down the generations – a process called "vertical inheritance" – bacteria can simply swap chunks of useful DNA with other microbes they meet. This means the microbes can gain new and useful biological traits in an instant. It's a strange phenomenon that biologists call "horizontal gene transfer."

The human factor has encouraged the spread of antibiotic resistance, too. Antibiotics are often misused in medicine and overused in livestock farming. This means that some bacteria have near-constant exposure to low-levels of antibiotics, which gives them even more opportunity to evolve ways to dodge the drugs.

Some observers fear that it's just a matter of time before superbugs that can withstand every antibiotic treatment available become widespread. They say we are on the cusp of a post-antibiotic era that could send medicine back to the Dark Ages.

But, as *Bald's Leechbook* reveals, perhaps medicine in the Dark Ages wasn't quite as bad as we might assume. This ancient medical text contains, among other things, a recipe for a treatment to tackle a common eye infection.

In 2015, Christina Lee at the University of Nottingham, UK, and her colleagues decided to follow the recipe.

The concoction – which is made by mixing leek, garlic and wine with salts extracted from a cow's digestive

Bald's Leechbook contains a recipe that can kill modern superbugs.

system – proved to be surprisingly potent. In tests it could kill MRSA, the notorious hospital superbug. The scientists suspect it's the way the ingredients combine that makes the therapy so effective: none of the ingredients individually seemed to be as powerful.

Bacteria were not formally discovered until the seventeenth century, but evidently Anglo-Saxon medics were aware of the infections they caused and effective ways to treat them.

Ancient texts are not our only weapon in the fight against superbugs. The specter of antibiotic resistance is prompting a hunt for new antibiotics – and it's

GO WIDER
SIX DEGREES OF SCIENCE

For more on...
Microbes behaving oddly:

The microbes that eat and poo electricity
Gene editing just got serious

 Even a handful of soil can contain
microbes with the power to kill bacteria.

astonishing where that hunt is taking scientists.

Some have begun scouring Earth's most remote environments. They are raiding the genomes of strange microbes that live below the ocean floor. In 2012, a fungus closely related to the one that gave us penicillin was found living here.

Others are sifting through the DNA in soil. Surprisingly, very few species of microbes can be grown and studied in the lab. However, by analyzing DNA in soil samples, biologists can identify new organisms that might have useful antibacterial properties. In 2015, Kim Lewis of Northeastern University in Boston, Massachusetts, and his colleagues found a brand new class of antibiotics in soil taken from a field in Maine – the first discovery of its kind in 30 years.

Perhaps the most surprising new antibiotic of all is one that Andreas Peschel at the University of Tübingen, Germany, and his colleagues found in 2016. It is produced by bacteria that live in human nostrils. These nasal bacteria produce a natural antibacterial chemical to fight off other bacteria that might want to take up residence there. Tests showed that the chemical is also effective against MRSA.

One solution to antibiotic resistance was under, or rather in, our noses all the time – and Peschel has a hunch that other brand new antibiotics will be found inside the human body as scientists learn more about its internal workings.

THE GENES THAT LIGHT UP AFTER DEATH

Breathing stops. The pulse weakens and ebbs away. Body temperature drops. Rigor mortis sets in. And then, hours or even days after the moment of death, something astonishing happens. Hundreds of genes spark into life.

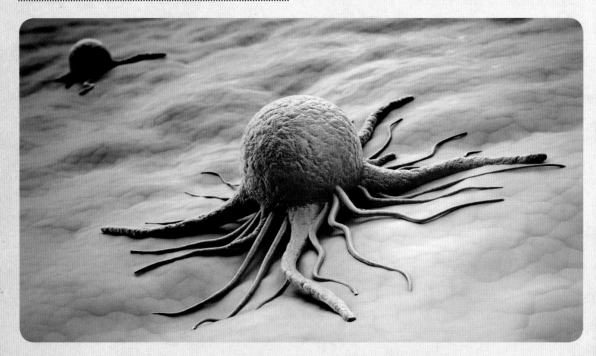

You might think that scientists have learned all there is to know about the way the human body breaks down after death. In 2016, they discovered that there are almost certainly still mysteries to unravel.

Microbiologist Peter Noble at the University of Washington in Seattle, and his colleagues, decided to explore exactly how the complicated genetic machinery inside animal cells "winds down" once life is over. To do so, the scientists studied two lab animals of choice for genetics research: mice and zebrafish. Noble and his

Some of the genes associated with cancer might become more active after we die.

colleagues took tissue samples from the animals every few hours for the four days following their death. Then they analyzed each sample for signs of genetic activity.

What they found came as a surprise. Not only did genetic activity in the two species continue for at least 48 hours after death, some genes actually became more active in the dead bodies as time passed.

Genes associated with tissue inflammation, the cellular response to stress, and the immune system were modestly active in the hours immediately following death. They then jumped to a higher level of activity – sometimes a full day or more after the animal had died.

The same curious phenomenon applied in the case of genes that are usually associated with cancer.

Noble and his colleagues are still trying to explain these strange findings. It's possible, for instance, that the genes might kick into action as part of some desperate and futile attempt to heal the body – even though the body is clearly far beyond help once death has occurred.

The work appeals to a morbid curiosity that many people may secretly harbor, which probably helps to explain why it made the science news headlines. But there is potentially real practical value to the science, too.

The discovery that some cancer genes burst into life long after death might be particularly significant.

Medical scientists are all too aware that people who receive organs from dead donors are more likely to experience poor health – including a raised risk of cancer – after the transplant operation. No one is quite sure why, although many think the recipient's health probably suffers because of the powerful drugs they must take to suppress their immune system and avoid it attacking and rejecting the transplanted tissue.

The new research suggests something else might be going on as well. Assuming that human genes switch on after death in the same way that mouse and fish genes do, the organs taken from a dead donor may contain a host of unusually active cancer genes. Once transplanted into a healthy, living recipient, those cancer genes could potentially remain active.

Crudely speaking, it's possible that an organ recipient might "catch" cancer from their dead donor.

It's an intriguing idea. If confirmed, it might give added

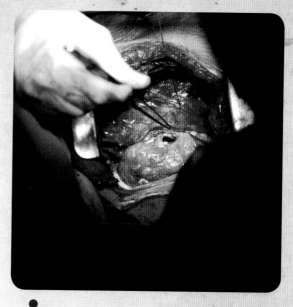

Transplant organs from dead donors might carry an unexpected cancer risk.

impetus to ongoing research efforts to build synthetic versions of human organs for transplant procedures.

The "dead gene" discoveries have other implications. Forensic scientists are always on the lookout for new ways to refine their estimates of the time of death in investigations. If it turns out there is a genetic dance of death that plays out in a predictable way and at a predictable pace in the decaying human body, the lives of forensic scientists everywhere might get that little bit easier.

Noble and his colleagues' initial work in dead zebrafish and mice suggests the approach has promise: when certain groups of genes were analyzed together, their combined activity did provide an accurate time of death prediction.

HOW WE CAN WORRY OURSELVES TO DEATH

In the 1970s, one man received awful news from his doctors. He had terminal liver cancer and would be dead in months. And, sadly, he was.

....

Then, and only then, did doctors discover something completely unexpected. An autopsy revealed that the man's liver tumor was tiny – certainly not large enough to prove deadly. The man hadn't died of cancer. He had died from the belief he was dying of cancer.

He was the victim of a strange phenomenon that scientists call the nocebo response.

The nocebo effect has been called the evil twin of the famous placebo effect. Where placebos seem to make people get better even though they contain no active medicine, a nocebo response can do the opposite. The mere suggestion of impending ill health can itself lead to a real deterioration in someone's condition.

The placebo effect is very well studied – the nocebo effect isn't. But in 2012, Paul Enck at the Tübingen University Hospital, Germany, and his colleagues compiled a list of as many reported nocebo cases as they could find. The nocebo effect is surprisingly common, they concluded.

It even has the potential to kill, as the example of the man who believed he had terminal liver cancer shows.

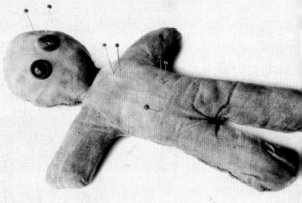

Voodoo medicine might tap into the nocebo effect.

(Below) Should doctors withhold information about a drug's potential side effects?

Another case that Enck and his colleagues found concerns a man in his twenties who was taking part in a clinical trial for a new antidepressant. The man used the experimental pills to take an overdose. Fortunately,

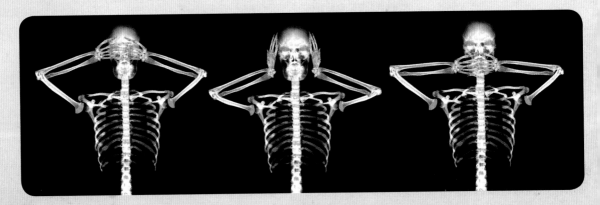

he was discovered, but not before his blood pressure had dropped to a dangerously low level. He was rushed to hospital to undergo urgent medical treatment.

A little later, the man was informed that the pills he had overdosed on actually contained harmless sugar rather than an experimental new antidepressant. New drugs' trials typically see half of the volunteers secretly assigned to a "control group" and half given the real drug. Volunteers don't know to which group they belong, but the man clearly suspected he was taking the real drug – and so his body reacted as if he had taken a dangerous overdose. Once it had been explained to him that his pills were harmless, even when taken in large numbers, his condition spontaneously improved.

Scientists are still trying to work out why the nocebo effect exists. At one level it seems tied to a deep sense of faith. Most people in developed societies have enormous trust in modern medicine, so when a doctor gives them a poor prognosis – even if it's incorrect – it has a powerful effect.

The same thing almost certainly goes on in other societies, too. To someone who believes firmly in the power of voodoo practitioners, for instance, being told that they have been cursed by one could well lead to a real deterioration in their health. Voodoo hexes might work because of the nocebo effect.

But there is probably more to the nocebo effect than a strong trust in doctors, not least given some evidence that a similar response can exist in animals – which generally don't have any faith to put in medicine.

Some nocebo responses might be tied up with anxiety. Simply being told something will hurt makes us anxious, and this seems to enhance the activity of pain receptors. The pain we sense really is worse as a result. Animals, too, have a stronger reaction to pain when they are in an anxious state.

All of these investigations into the nocebo response leave medical doctors facing an ethical dilemma.

Doctors often pledge to "abstain from all intentional wrong-doing and harm" – an ancient tradition that began with the famous Hippocratic oath. But, as the case of the cancer patient in the 1970s shows, a doctor's words can have a powerful influence on a patient's well-being.

In their 2012 paper, Enck and his colleagues suggested that in some cases a doctor might consider making an unusual request of patients with serious conditions. The medic might ask for permission to avoid discussing

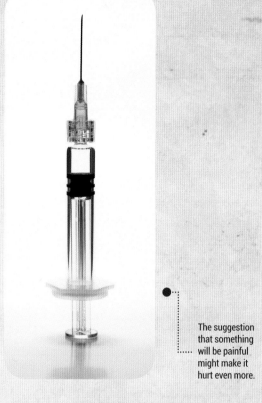

The suggestion that something will be painful might make it hurt even more.

possible side effects of a treatment – just in case the mere discussion of such symptoms is enough to trigger their occurrence. Treating conditions like cancer is challenging enough without the patient's own body unconsciously acting to make the therapy even less likely to succeed.

When it comes to the potential downsides of a medical treatment, ignorance might just be bliss.

GO WIDER

SIX DEGREES OF SCIENCE

For more on…

The placebo effect:
The power of the placebo

The story of cancer:
Why cancer is like a selfish animal

THE FAT THAT MAKES YOU THIN

There's a health condition spreading rapidly through the global population. It affects young and old, rich and poor, male and female – and some evidence suggests it's contagious. The obesity pandemic is one of the most significant challenges facing the medical world. By 2009, a new and very surprising weapon in the war against obesity was beginning to emerge. Body fat itself might be a potent anti-obesity tool.

The availability of affordable processed food probably does most to explain the rise of obesity in almost all countries. Other factors may be at work too, though. There are, for instance, anecdotal reports that people undergoing a so-called fecal transplant to treat a gut infection have "caught" obesity from their overweight donor.

Whatever the root cause, the consequences are generally the same. Obesity leaves people with excessive deposits of white adipose tissue, or more simply "white fat." This white fat offers the body a location in which to store excess energy that it can't immediately use.

But there is another and less abundant type of fat, called brown fat, that works in a different way. Brown fat is packed full of energy-burning structures. Here, crudely speaking, excess energy is burned rather than stored, kicking out heat in the process.

Biologists have known about the existence of brown fat for a very long time. Its heat-generating properties are crucial in mammals that hibernate through cold winter conditions. It was established long ago that humans have brown fat deposits, too. Babies rely on them to generate body heat because, unlike adults, babies can't shiver to keep warm in cold weather.

But the consensus until a few years ago was that humans lose most, perhaps even all of their brown fat by adulthood. A number of studies in 2009 challenged that consensus. Scientists found that, unexpectedly, human adults do have brown fat deposits – generally around the chest and neck.

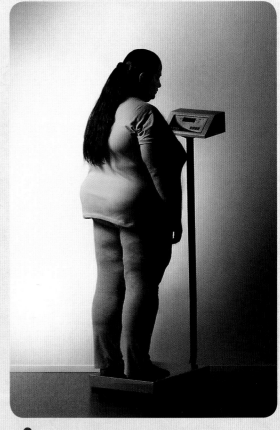

The obesity pandemic is growing worse by the year.

Babies carry deposits of brown fat to keep warm in cold conditions.

"feed" on white fat deposits in its hunger for calories to burn. This could, in principle, help people to gradually erode deposits of excess body fat without the need to exercise.

There's a downside, though. Carpentier found these effects only when his male volunteers donned a special suit that pumped cool 18 °C (64.4 °F) water over their bodies. Brown fat becomes active at relatively low temperatures – conditions that some people might view as intolerably cold.

There's another problem. Most adults don't carry enough brown fat tissue to eat through really excessive deposits of white fat – although indigenous communities in the Arctic seem to carry more brown fat than adults in other regions.

Ideally, scientists will have to come up with a way to encourage the growth of brown fat in adults – or better still, find ways to make white fat deposits "turn brown."

There are encouraging signs that this might actually be possible. In 2012, Bruce Spiegelman at Harvard Medical School and his colleagues confirmed that mammals carry a third type of fat – called "beige fat" – that shares some of the energy-burning properties of brown fat.

Some evidence suggests that white fat can be coaxed into becoming energy-burning beige fat. One way again involves exposure to cold conditions. Another method is to adopt a diet extremely low in calories – diets that have also been linked to a prolonged lifespan in mammals. Arguably, of course, if people who are obese are willing to take on one of these extreme "calorie-restricted diets," they won't remain obese for long in any case.

If biologists can discover a way to stimulate the white-to-beige conversion process without cold exposure or calorie restriction – and some teams are working on doing just that – then the result could be something close to the ultimate obesity treatment: a drug that encourages body fat to burn itself into oblivion without the need for exercise, cold exposure or any discomfort.

Since then, scientists have been busy working out whether brown fat can help adults burn through calories and, in theory, prevent weight gain.

The good news is that it can. In 2012, for instance, a team led by André Carpentier at the University of Sherbrooke in Quebec, Canada, found that brown fat burned through calories in adult male volunteers even when they were inactive. What's more, there were tantalizing signs that brown fat might even begin to

GO WIDER

SIX DEGREES OF SCIENCE

For more on...
Why Arctic communities have more brown fat:

 Extinct humans found in our DNA

Prolonging human lifespans:

Is this the secret to eternal youth?

THE LIFE-SAVING POWER OF EXCREMENT

A new class of medical donor has emerged in the last decade. It's not their blood that they have decided to share with the sick. It's not even their internal organs. It's their excrement. The gut microbes that get caught up in our feces have the potential to cure serious and potentially life threatening infections.

Scientists have known for decades that the human body plays host to a range of microbes. But until about 15 years ago, those microbes were frustratingly difficult to study: surprisingly, the vast majority of microbes on Earth – including those in our bodies – refuse to grow in Petri dishes in the lab, which makes the job of studying them very challenging.

Things have now changed. Geneticists can use new state-of-the-art lab equipment to sequence all of the DNA in a sample. Doing so gives the scientists an idea of the diversity of microbes present in a particular environment.

Applying the technology to samples collected from various parts of the human body revealed an unexpectedly diverse and abundant community of bacteria. In fact, geneticists concluded that the microbial cells in and on the average human body might outnumber actual human cells by 10 to one – although in 2016 this estimate was scaled down: our bodies probably carry "only" about one microbe for every human cell, or about 40 trillion microbes in total.

The discovery of this shadow self we all carry around has begun to have an impact on all sorts of scientific investigations.

Forensic scientists have found evidence that the microbial community inside the human body – the "microbiome" – changes in a steady and predictable way after death, offering a new way to establish time of death in forensic investigations.

NASA, meanwhile, is eager to understand how the microbiome changes during space flight and whether this should influence the medical treatment astronauts receive – something particularly important to bear in mind ahead of proposed manned trips to Mars. In 2015, the space

Living at zero gravity for a year changed the bacteria in astronaut Scott Kelly's gut.

agency monitored astronaut Scott Kelly's gut microbiome during a year-long stint on the International Space Station (ISS), while simultaneously monitoring the gut microbiome of his identical twin brother, Mark, back on Earth.

Medical treatments on Earth have as much to gain from microbiome studies, particularly in light of research led by Alex Khoruts at the University of Minnesota Medical School in Minneapolis. Khoruts showed it is possible to modify the human gut microbiome, which means if – for whatever reason – someone's gut microbiome becomes "sick," medics can help it recover.

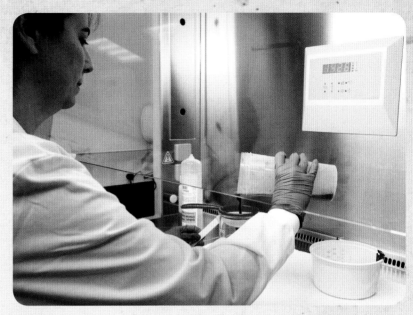

The discovery led to a surge of interest in fecal transplants. Some of the bacteria in the human gut are expelled, alive, in human feces. Khoruts and other medics treat the feces from healthy donors and then transplant them into patients with gut-related health conditions. The hope is that the beneficial bacteria in the transplant will establish themselves in the donor's gut and nudge it into a healthier state.

The fecal transplant procedure is not for the squeamish. Sometimes the filtered fecal matter is fed into the recipient's stomach via a nasal tube. Other medics prefer taking an alternative route, administering the transplant as an enema.

But the results can be spectacular. People with life-threatening bacterial infections that have developed resistance to almost all antibiotics have made full recoveries thanks to a fecal transplant. Fecal matter even seems to help relieve some of the symptoms of Parkinson's disease.

In fact, human excrement has so much potential to improve health that US authorities began regulating it as a drug in 2013.

However, as with many drugs, fecal matter may come with side effects. Some recipients of fecal matter from an overweight donor have become obese following the transplant. They are adamant that they have "caught" their obesity from their donors. There is growing acceptance that gut bacteria can impact someone's

Excrement must be carefully treated before it is "transplanted" into a donor.

weight. Even so, the idea that fecal transplants could "transmit" obesity is still generally regarded as unproved at present.

Nevertheless, it's better to be safe than sorry, and changes have been made to the screening process used to decide who is a suitable fecal donor. Donating feces is rapidly becoming a serious business.

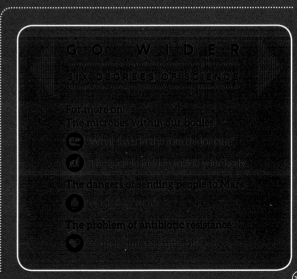

GO WIDER
SIX DEGREES OF SCIENCE

For more on...
The microbes within our bodies:
What lives in the fourth domain?
The secrets hidden inside your body

The dangers of sending people to Mars:
Not too Mars

The problem of antibiotic resistance:
Catch the time for antibiotics

HUMAN ORGANS ON DEMAND

The world is in desperate need of more organ donors. Reports suggest that several hundred patients in the UK die each year while waiting for an organ transplant. In the US, those annual deaths number in the thousands, according to the Organ Procurement and Transplantation Network, a public-private partnership.

Campaigns to encourage people to donate their organs after death might help deal with the crisis. But biotechnologists are working on another way to help people who need a replacement liver, kidney or heart. Soon, doctors might simply print patients a new organ on demand.

....

Not long ago, 3D printers were seen as a novelty. Today the technology has spawned a global multi-billion dollar industry. Most of the hearing aids sold worldwide are reportedly 3D printed. One company – Local Motors, based in Phoenix, Arizona – has plans to revolutionize the car industry by 3D printing vehicles in the consumer's hometown. Manufacturers are even turning to 3D printing to churn out cheap unmanned aerial vehicles, or drones – another fast-growing and versatile industry that can meet all sorts of scientific needs.

But of all the uses so far found for 3D printing, it's "bioprinting" of human organs that is perhaps the most exciting – and potentially the most useful.

The idea builds on existing successes in tissue engineering. For about 15 years, scientists have been able to make replicas of simple body parts using special biodegradable plastics. These plastic structures then act as a scaffold on which human cells can grow and replicate.

Over time, the human cells grow into strong tissue, and the polymer degrades and falls away. The end result: a healthy body part ready for transplant.

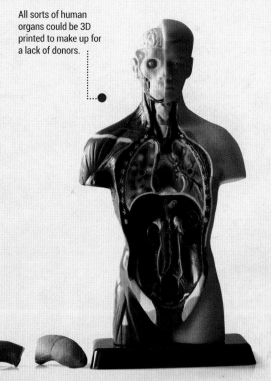

All sorts of human organs could be 3D printed to make up for a lack of donors.

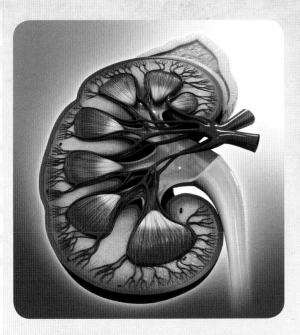

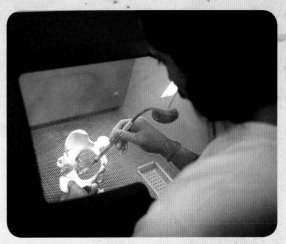

(Left) The internal structure of the kidney will be difficult to replicate with a 3D printer.

(Above) Medics are already using 3D printers to produce models of the human skeleton.

What makes this sort of biotechnology particularly compelling is that the body parts can be built using cells take from the intended recipient. This means that the lab-grown structure is effectively already part of the recipient's body even before it is transplanted, making it less likely to be rejected by their immune system – which is one important advantage that custom-built organs have over organs taken from dead donors.

Using this approach, a team led by Anthony Atala at the Wake Forest Institute for Regenerative Medicine in Winston-Salem, North Carolina, has made bladders and urethras – and successfully transplanted them into people with medical conditions.

Most recently, in 2014, Atala's team added vaginas to the list of body parts that they could grow in the lab and transplant.

But all of these body parts are relatively simple – basically just tubes or balloons. They are so simple that Atala's team actually hand-stitched the plastic scaffolds onto which the human cells were later added.

This is where 3D printers come in. A 3D printer can build very complicated structures – and it does so in a fraction of the time it takes to hand-stitch a body part scaffold. Atala hopes that eventually it will be possible to 3D print complicated scaffolds and use them to grow organs like the kidney or liver.

His team has even produced conceptual models: kidney-shaped structures built by a 3D printer.

There's a long way to go before scientists can 3D print all of the delicate vessels found inside a real kidney, though, and an even longer road to travel before such complicated scaffolds can be seeded with human cells and turned into real organs.

Until those significant challenges can be met, the kidney-shaped structures that Atala and his colleagues have built will remain tantalizing – but functionless – signs of what might eventually be possible.

If Atala and his team do meet their ultimate target, though, a 3D printer might one day save your life.

GO WIDER

SIX DEGREES OF SCIENCE

For more on...
Scientific uses for drones:
The drone that controls the weather

The health effects of organ transplants:
The genes that light up after death

HOW BEER COULD HELP CURE MALARIA

Apparently, the Bulgarians were late to appreciate the art of brewing. Traditionally a wine-drinking country, it wasn't until the nineteenth century that foreigners brought in beer.

When it comes to the latest revolution in brewing, though, Bulgaria is right at the forefront. A few years ago, a chemical company in the country filled its brewers' vats with a brand new form of yeast. This yeast doesn't ferment sugar to produce alcohol – it churns out a key ingredient in the most powerful anti-malaria drug on the medical market.

Malaria casts a great shadow over the global human population. According to the World Health Organization (WHO), 3.2 billion people – almost half the humans on Earth – are at risk of catching the mosquito-borne disease. WHO estimates that nearly half a million people died of the condition in 2015 alone.

But these figures don't tell the whole story. WHO also says that measures to control the disease and prevent its spread have helped reduce the malaria death rate some 60 percent since the year 2000.

Right at the forefront of those successful measures is a drug called artemisinin, which is produced naturally in the leaves of sweet wormwood, a plant that grows in China.

In the early 2000s, pharmaceutical companies began using sweet wormwood to manufacture artemisinin treatments. At about the same time, WHO recommended artemisinin should be part of the standard treatment for malaria and its production boomed.

The artemisinin for those treatments came from farmers growing sweet wormwood in China and other parts of the world. They struggled to meet the demand and by 2004 there was a shortfall in production. It was a problem in need of a solution.

At about the same time Jay Keasling, a biologist at the University of California in Berkeley, had devised a solution in need of a problem. Keasling is one of a new wave of "synthetic biologists" who make wholesale

Yeast fermentation is no longer just about generating alcohol.

changes to the genes inside simple microbes to completely alter their biology. Synthetic biology is essentially an extreme version of genetic modification.

Keasling wanted to turn simple brewer's yeast into microbes that churn out pharmaceutically useful chemicals – but he couldn't decide which chemical to aim for. The artemisinin shortfall gave him his target.

By 2006, Keasling's team had engineered yeast that could churn out a substance called artemisinic acid, which is then relatively easy to convert into artemisinin. Keasling began working with a pharmaceutical company, Sanofi, to improve production yields. By 2013, those commercial fermenters in Bulgaria were ready to fire up.

In the years that followed, however, the fermenters have fired down again.

The problem is not with Keasling's science or with synthetic biology: it's with the commercial world. While Keasling was developing his yeast, the artemisinin market became hugely volatile. Following the 2004 shortfall in production, more farmers began growing sweet wormwood, recognizing a commercial opportunity. Predictably, this quickly led to overproduction and, by 2007, the price of artemisinin had slumped.

Organizations sprang up to try to smooth the supply and demand problems – and they were largely successful. By the time the synthetic yeast was ready to feed into the commercial artemisinin market, that market was far more stable and shortfalls in production far less likely.

This meant that the synthetic source of artemisinin – which was only ever intended as an emergency supply in case of shortfalls – was no longer really necessary.

This is symptomatic of a general problem facing synthetic biology. The science is now at a stage where it's possible to turn yeast into tiny factories that pump out all sorts of useful and valuable chemicals – the main components in saffron, the artificial sweetener stevia, or vanilla, to name just three. But at a social level, there is hostility to these projects. Critics say life is tough enough for farmers – particularly those living in relatively impoverished parts of the world – without the added problem of having to compete with synthetic biology.

There's no doubt that synthetic biologists can make astonishingly impressive and potentially useful modifications to microbes. But, in a sense, the technology is still a solution looking for a problem.

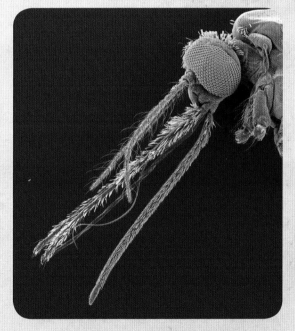

(Above) Sweet wormwood is the traditional source of artemisinin.

(Right) Malaria-carrying mosquitoes remain a major health problem.

GO WIDER

SIX DEGREES
OF SCIENCE

For more on...
Extreme genetic modification:
What synthetic life reveals
about the living world

The challenges that farming brought:
Was this humanity's
biggest mistake?

IS THIS THE SECRET TO ETERNAL YOUTH?

In June 2015, the battle against aging got serious. The United States Food and Drug Administration (FDA) approved a clinical trial to explore whether a drug already widely used to treat type 2 diabetes could slow the human aging process.

It was reportedly the first time the administration had recognized aging – rather than a specific disease – as a condition worth targeting with drugs.

Probably as long as people have been aware of their own mortality, there have been those with a burning desire to cheat death. This is despite the fact that myths and other stories going back millennia warn of the dangers that can come with immortality. Those warnings find an echo in the realm of contemporary science: human cells can become "immortal" – but they do so by mutating into cancer cells with the power to cut short human life.

Perhaps it's wiser instead to aim to delay death rather than avoid it. People today already benefit from average lifespans that are greater than they were in prehistory – but there's always room to boost those averages a little further. To that end, many people will be eagerly awaiting results from the FDA-approved study into metformin.

Chemists first described metformin in the 1920s. By the 1950s it was being used to treat people with diabetes.

Humans have been searching for the fountain of youth for millennia.

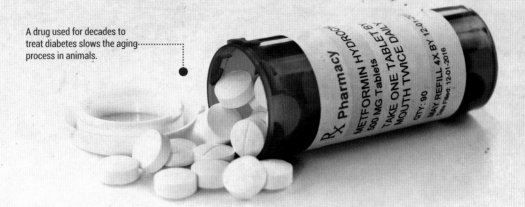

A drug used for decades to treat diabetes slows the aging process in animals.

Metformin is a cheap and effective way to reduce blood sugar levels. So far, so normal.

However, since about 2005, scientists have begun to notice something odd about metformin. Lab animals – worms and mice – age more slowly than expected when they are given the drug. They live longer, too, and they appear to remain healthier than they otherwise should. For instance, if the animals develop cancer, metformin seems to help slow tumor growth.

Humans may benefit from some of these effects, too. Scientists studying people who were taking metformin for diabetes noticed that their test subjects tended to show a lower than expected chance of developing heart disease, and a lower risk of developing cancer.

The question scientists want to answer next is a simple one. Can the drug slow down the aging process in humans and delay death?

A team led by physician Nir Barzilai at the Albert Einstein College of Medicine in New York City hopes to find out through its "Targeting Aging with Metformin" (TAME) clinical trial. Barzilai is still securing funding for the study, which will involve giving some people in their 70s with conditions like cancer or heart disease metformin. Others in a control group will receive a placebo. Then the researchers will simply watch to see what happens. They will monitor the volunteers in both groups to see whether or not the people taking metformin stay healthier for longer, and ultimately whether they have longer lifespans.

We've been here before, of course. All manner of scientific studies have apparently pointed the way towards the fabled "fountain of youth." Not long ago it was a chemical found in grape skins – and in red wine – that some saw as the key to a longer life. The chemical in question, resveratrol, does appear to affect lifespans in some species. But, disappointingly, not in humans.

Extreme diets are also seen as a route towards a longer life. Scientists have known for decades that lab animals fed on very limited amounts of food live longer – but exactly why is unclear. A few committed people have begun eating "calorie-restricted" diets themselves, and some scientific studies hint that the diets do make a difference. The internal organs of these people are aging at a slower rate than normal.

But many people may have hopes of living well beyond their 100th birthday without sacrificing their love for rich and hearty food. These individuals might want to keep an eye on the TAME trial. It's just possible that an effective anti-aging pill is on the horizon.

GO WIDER
SIX DEGREES OF SCIENCE

For more on...
Dodging the grim reaper
The secrets of the immortal
Could a frozen squirrel help humans cheat death?

The differences between modern and prehistoric people
Why our brains are shrinking

Calorie restricted diets
The fat that makes you thin

HOW A MAN'S NOSE HELPED HIM WALK AGAIN

They say you never forget how to ride a bike. In a sense, then, it's not surprising that Darek Fidyka found it relatively easy to remember how to pedal after several years out of the saddle. But news that he was cycling again still made international headlines in March 2016. That's because a knife attack in 2010 had severed Fidyka's spinal cord and robbed him of all feeling from the chest down.

Fidyka's recovery – and reports that other people are regaining feeling in previously paralyzed limbs – shows the astonishing progress that medical scientists are making in the treatment of spinal cord injuries.

....

The nervous system is incredibly complex – and the spinal cord is the jewel in the crown of that complexity. It's little wonder that neuroscientists have traditionally faced enormous difficulties in trying to treat severe spinal cord injuries.

Injuries to the spinal cord are extraordinarily difficult to treat.

Until very recently, in fact, no patients had benefitted from spinal regeneration therapy.

This is now beginning to change, as Fidyka's case demonstrates. His treatment involved olfactory ensheathing cells (OECs), a type of cell that is found in the olfactory bulbs located just above and behind the nose. These cells have evolved to repair nerves involved in our sense of smell.

In 1997, a research team led by Geoff Raisman at University College London in the UK found that OECs could repair nerves in the spinal cord, too. Simply injecting OECs into the site of a spinal cord injury helped reverse some of the injury's effects in rats.

By 2012, Raisman and his colleagues were ready to trial the procedure in a human volunteer – Fidyka. He went under the knife to have one of his olfactory bulbs removed, from which OECs were extracted and cultured in the lab. When the researchers had enough OECs, they injected them into Fidyka's spine immediately above and below his injury. They also placed strips of nerve tissue across the injury to act as cellular bridges across which the OECs could perform their nerve-repairing feat.

Just three months later, Fidyka was already reporting that the muscles in his thighs were gaining strength. By 2014, a couple of years after the surgery, he was able to walk with the aid of a supporting frame.

in his legs to
control he is
legs, too.
uel Nicolelis,
of Medicine
ing. He is
eating spinal
sses.
t can
een working
al cord
once

headset
ense of
signals are
generated
ty to learn to
elis's team call

al avatar
ld. Nicolelis's
robotic
ed legs that
rain signals
olunteers can
about moving

nteers to walk
robot motors
vork.

is initial set

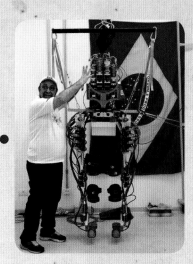

Miguel Nicolelis
has had astonishing
success treating
people with
spinal injuries.

of eight volunteers found that some feeling and muscle strength was returning to their paralyzed limbs. Seven of the eight have even been "upgraded" from complete paralysis to partial paralysis.

At this point, none of the eight can walk without help, but regaining even weak muscle control after severe spinal injury is remarkable.

Exactly how or why Nicolelis's approach has restored some feeling to his patients is still a mystery. But the work offers another glimmer of hope that some forms of severe paralysis might be treatable.

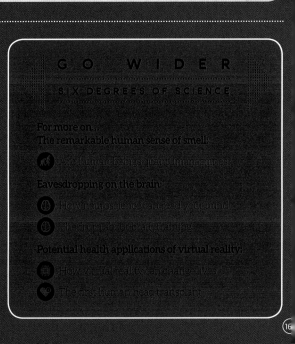

G O   W I D E R

SIX DEGREES OF SCIENCE

For more on...
The remarkable human sense of smell:

Are human beings quantum machines?

Eavesdropping on the brain:

How neuroscience can read your mind

The truth about brain training

Potential health applications of virtual reality:

How virtual reality can change lives

The first human head transplant

THE FIRST HUMAN HEAD TRANSPLANT

It's not often that stories in the popular scientific press carry warnings of graphic and disturbing images. However, they did in January 2016. Neuroscientists revealed a gruesome image showing the decapitated head of one monkey stitched on to the headless body of a second. The research was the latest step down a hugely controversial avenue of medical research: the science of human head transplants.

Researchers are debating the pros and cons of the ultimate transplant operation.

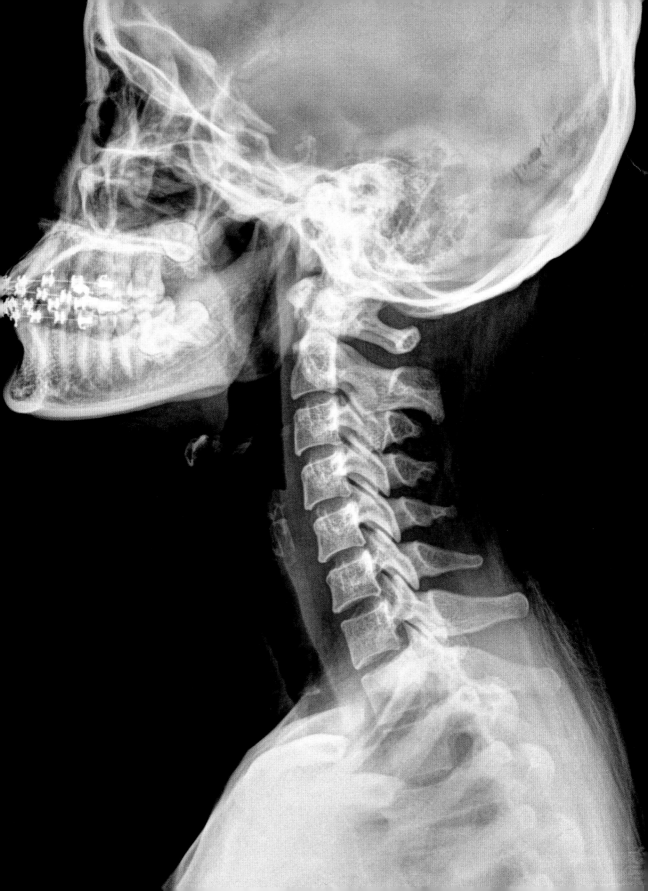

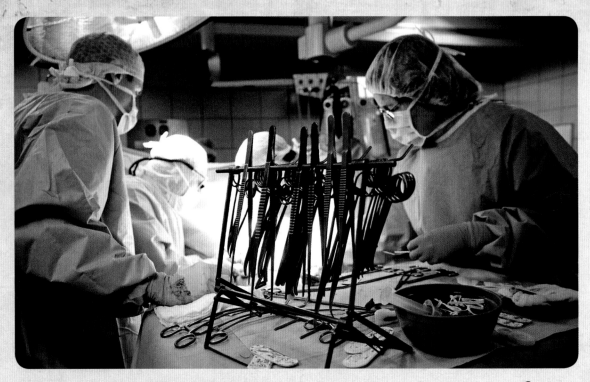

Interest in the idea of human head transplants spiked in 2015 when an Italian surgeon, Sergio Canavero, announced to the media that he planned to perform the procedure within two years.

Canavero thinks that the surgery can help people with conditions that rob them of control over their body. Such a person could have their head carefully removed from their diseased body and then transplanted onto the healthy body of a recently dead donor.

But the idea has been met with near universal criticism by surgeons, scientists and ethicists, who say the science behind the idea is far from proved.

There are obvious reasons for their concern.

The brain is a very sensitive organ. If it is deprived of oxygen for even a few minutes, irreparable brain damage occurs. The spinal cord is a complex structure. Once it is cut, the chances of completely healing the injury seem remote – even without the added complication of having to connect the spinal cord in the neck of one individual to the cord in the body of a second person. Correctly stitching together all of the muscles that pass through the neck will be a massive challenge too.

Even if the surgical procedure passes off successfully, there is no way of knowing how someone would adjust

Transplants that are now considered routine were once very controversial.

to life attached to a completely foreign body. Their whole sense of identity might change in an unpredictable and potentially very distressing way.

Canavero is confident that all of the obstacles can be cleared and that all of the problems have solutions. Perhaps some of them do.

Paralysis, for instance, does not look as hopelessly irreversible as it did a few years ago. There are tantalizing signs that medical research can help some people with severe spinal injuries. A few volunteers have undergone pioneering medical procedures that seem to have restored some degree of feeling and muscle control to previously paralyzed limbs.

The risk of brain damage, meanwhile, could be controlled to some degree by cooling the body before the operation. Canavero says this is what the surgeons who performed the controversial monkey head transplant procedure did. It's conceivable that scientific investigations into animals that allow their bodies to cool – even to freeze – in winter might point to new techniques that could help safely cool a human brain before a head transplant.

Sergio Canavero has plans to conduct the first human head transplant operation.

The psychological problems that someone might encounter as they adapt to life in a new body are virtually impossible to predict. But advances in virtual reality technology are allowing psychologists to explore the concept of self in new ways. Volunteers can "swap" gender, ethnicity, age and more in virtual reality. Such experiments might at least offer a flavor of the problems that someone might face after a successful head transplant.

It's important to stress that most scientists view the idea of head transplants with severe skepticism – they say that far more scientific research is needed before anyone can even consider the prospect. And if the science ever reaches the point where such a surgical procedure is possible, the ethics of the operation will need to be carefully considered and debated.

But, as some media commentators have pointed out, transplant history has faced these controversies in the past.

Richard Lawler became the first surgeon to carry out a successful kidney transplant in 1950. Some of his contemporaries accused him of playing God – and

Lawler later recalled being ostracized by other surgeons, even those he had considered to be close friends. Today, few people consider kidney transplants to be controversial: is it conceivable that the same may one day be true of head transplants?

GO WIDER

SIX DEGREES OF SCIENCE

For more on...
Reversing some forms of paralysis:
How a man's nose helped him walk again

Cooling biological tissues safely:
Could a frozen squirrel help humans cheat death?

Body-swapping in virtual reality:
How virtual reality can change lives

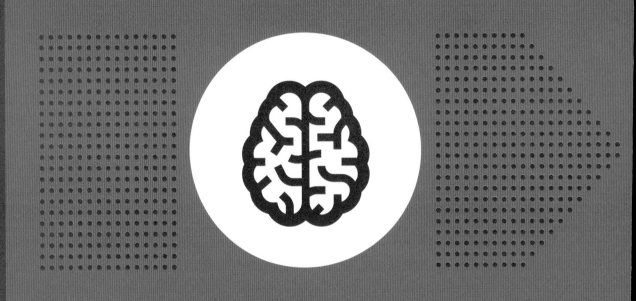

CHAPTER SEVEN

THE BRAIN
& HUMAN BEHAVIOR

THE PARASITE THAT MAY BE MANIPULATING YOUR BEHAVIOR

We like to think that we rule our own minds. But for one-third of the global human population, things might be a little more complicated. These people carry a tiny parasite in their brain that seems to have the power to tinker with their behavior. This parasite may even encourage people to take life-threatening risks.

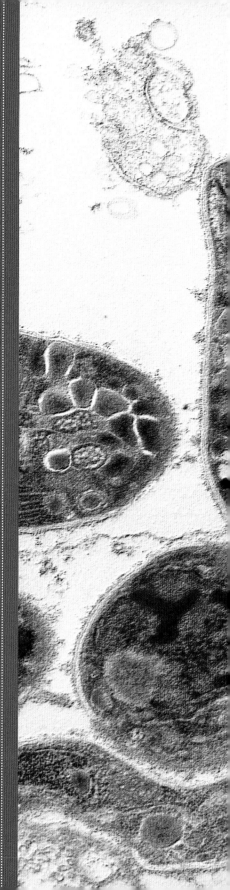

Toxoplasma seems to have the power to turn people into risk takers.

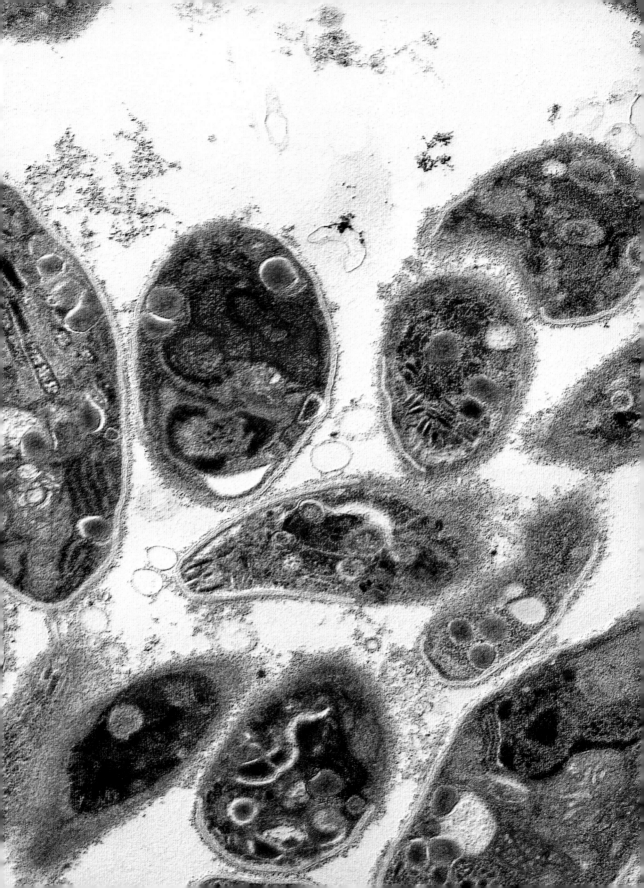

Toxoplasma is a single-celled microbe with a strange sex life. Although it can live inside all sorts of birds and mammals – including humans – it reproduces only inside cats. Understandably, then, it's in the parasite's interests to find a feline.

About 20 years ago, scientists discovered that the parasite has developed ways to speed up that journey into a cat.

A research team including Joanne Webster, now at Imperial College London in the UK, reported that rodents infected with *Toxoplasma* become unusually active and bold, which makes them an easier target for predators such as cats. By the year 2000, Webster's team had discovered something even more remarkable. The rodents will actually go out of their way to find a feline predator: rats with a *Toxoplasma* infection are attracted to the pungent odor of cat urine.

A biologist in the Czech Republic read about these discoveries with interest. Jaroslav Flegr at the Charles University in Prague suspected that *Toxoplasma* had the power to manipulate another type of brain: the human brain.

Flegr had already discovered that people infected with *Toxoplasma* have delayed reaction times, but it was a paper he and his colleagues published in 2002 that really attracted scientific attention. The researchers studied people who had caused traffic accidents on the streets of Prague. Both pedestrians and car drivers injured as a consequence of

their own actions were more than twice as likely as the average Prague resident to carry *Toxoplasma*.

The parasite seems to make people more casual about risk taking. Deep in prehistory, our ancestors were potential prey for big cats – perhaps *Toxoplasma* makes people take risks because once upon a time that increased the parasite's chances of getting into a cat.

A 2016 study backed up the idea. A team led by Clémence Poirotte at the Centre d'Ecologie Fonctionnelle et Evolutive in Montpellier, France, discovered that our closest living relative – the chimpanzee – loses its fear of leopard urine and is actually attracted to the scent upon infection by *Toxoplasma*. Leopards are the chimpanzee's only natural predator.

Somewhat alarmingly, it's all too easy for humans to pick up *Toxoplasma*. Infection can come from eating meat

Rats lose their fear of cats if they are infected with *Toxoplasma*.

GO WIDER
SIX DEGREES OF SCIENCE

For more on...
Surprising ways animals can be manipulated:
 When food bites back

that hasn't been fully cooked. Even simply changing a pet cat's litter tray can lead to exposure. In most people, the symptoms of initial infection are subtle – it can be mistaken for mild flu. The human immune system fights back, but *Toxoplasma* responds by migrating into certain tissues, including brain tissue, and entering a dormant phase. It's then that the parasite begins its nefarious mind manipulation.

Of course, no one likes the idea of their behavior being controlled by a foreign agent – particularly when that agent is a single-celled microbe. For neuroscientists, though, there's some grudging admiration for *Toxoplasma*. They are getting increasingly adept at understanding – and even manipulating – brain activity, but studying parasites like *Toxoplasma* might help them finesse their skills.

So how exactly does *Toxoplasma* meddle with our minds?

There's still no clear consensus, but there are clues. A study published in 2009 by Glenn McConkey at the University of Leeds, UK and his colleagues suggests the parasite might boost the production of a hormone called dopamine that helps modulate pleasure and fear in the human brain.

Other studies also point to a link with this hormone. A few years before McConkey's work was published, Webster and her colleagues had discovered that rats

with a *Toxoplasma* infection lose their death wish if they are given a drug called haloperidol – which inhibits dopamine production.

That might be a significant discovery. Haloperidol happens to be a drug often prescribed to treat schizophrenia.

In 2015, research led by E. Fuller Torrey at the Stanley Medical Research Institute in Kensington, Maryland, concluded that a *Toxoplasma* infection might double the risk of schizophrenia. Other research published in 2015, meanwhile, uncovered evidence that exposure to cats in childhood might be a risk factor for developing mental illness more generally.

Such findings are alarming but also, in a way, encouraging: they might be a first step towards finding effective treatments for some forms of schizophrenia and for other distressing conditions.

People at fault in a traffic accident seem to be more likely to be infected with the parasite.

HOW NEUROSCIENCE CAN READ YOUR MIND

The voice is a little indistinct, as if its owner is speaking underwater. The words are still reasonably clear, though – structure; company; doubt. But these three little words stand out among the many spoken every day, precisely because they weren't actually spoken. They were reconstructed inside a computer from analyzing brain activity.

Neuroscientists are venturing deeper and deeper into the brain in their efforts to understand – and potentially manipulate – the way it works. The results of their investigations are nothing short of astonishing.

By recording and analyzing patterns of human brain activity, these scientists can reconstruct what people are seeing and hearing. They can even eavesdrop on the internal narrator who gives voice to our thoughts.

These groundbreaking experiments begin with volunteers performing simple tasks. They might be asked to listen to a string of words, for instance, or stare at a sequence of very simple line drawings. As they do so, neuroscientists are using special scanners to record the way their brain responds. They note the precise pattern of brain activity that each word or image generates.

Eventually, the neuroscientists can predict exactly which word or image – from a set of several – someone is hearing or seeing just from the distinct pattern of brain activity it generates.

This is impressive enough, but a few years ago the neuroscientists decided they wanted to do something

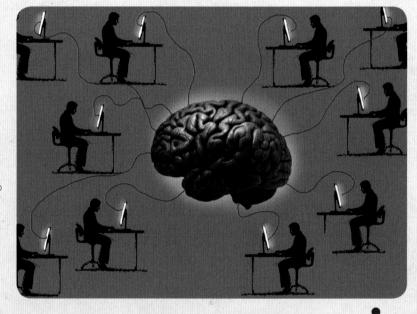

The inner workings of the brain are no longer as mysterious as they once were.

even more spectacular. Supposing a lab technician gave the volunteer a brand new image to look at: the neuroscientists wanted to be able to decode the pattern of brain activity it generated and reconstruct the image on their computer screen without ever looking at the image through their own eyes.

Incredible though it might seem, they succeeded.

Yukiyasu Kamitani at the ATR Computational

Neuroscience Laboratories in Kyoto, Japan, was among the first to achieve this breakthrough. In 2009, he and his colleagues asked volunteers to look at simple black-and-white images constructed using a 10 by 10 grid of pixels. Through careful analysis of a volunteer's brain activity, Kamitani's team found they could predict whether each of the 100 pixels in the grid was black or white in any given image. This meant that when the volunteer was presented with an entirely new image constructed using the pixel grid, the scientists could establish which pixels in the image were white and which were black. They could reconstruct the image just from brain activity.

By 2011, Jack Gallant at the University of California at Berkeley had gone even further. His research team used similar techniques to reconstruct fuzzy, low-resolution versions of YouTube video clips, simply by analyzing the brain activity of volunteers who were watching the real thing.

According to Gallant, eventually it might even be possible to hack into the "video stream" of someone's dreams and play it on screen, according to Gallant – although developments like that are almost certainly decades away.

Another neuroscientist – Brian Pasley, who also works at the University of California at Berkeley – is beginning to have similar sorts of success with sound. It was Pasley's team who, in 2012, found they could use computer software to reconstruct and play back the words someone was hearing – including "structure," "company" and "doubt" – just by decoding the pattern of brain activity that the words generated.

Neuroscientists listened in when people silently read Abraham Lincoln's famous words.

In 2014, Pasley took this idea to another level. He wondered whether we listen to our internal monologue in the same way that we listen to other people's voices. It turns out we do: he asked volunteers to silently read from the Gettysburg Address while their brain activity was being recorded. Gallant found that he could reconstruct individual words that people read but never actually spoke.

This sort of work might appall many – it seems like the ultimate privacy breach. But it's worth bearing in mind that brain decoding relies on willing volunteers who are prepared to sit still for long periods of time, viewing, hearing or reading the same material repeatedly while their brain activity is recorded and carefully analyzed.

It's also worth considering what this technology might mean to someone with a degenerative disease that has robbed them of the ability to speak. For some people, this technology could eventually lead to significant improvements in their quality of life.

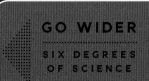

GO WIDER

SIX DEGREES
OF SCIENCE

For more on...
Analyzing brain activity:

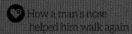 How a man's nose helped him walk again

The truth about brain training

Brain manipulation:

 The parasite that may be manipulating your behavior

CAN SCIENCE HACK YOUR SLEEPING MIND?

Something was different. On the previous day, the mice had wandered around the octagonal arena more or less at random. Now, they headed directly towards a particular corner and waited there, in apparent anticipation.

There was a logical explanation for their change in behavior. Overnight, scientists had hacked into the rodents' dreams and manipulated their brains. They had effectively created a memory that the mice had not made themselves.

The scientists had convinced the animals that they had received a reward in one corner of the arena. Upon finding themselves back in the arena, then, the mice made a beeline to that corner in the hope of receiving a reward for doing so.

Neuroscientists have suspected for some time that sleep – and in particular dreams – play an important role in memory formation. Even so, the idea that researchers can eavesdrop on the dreaming process and manipulate memories seems far-fetched: the stuff of Hollywood movies, not real scientific investigation. One study in 2015 changed that.

Karim Benchenane's team at the French National Center for Scientific Research in Paris gave an impressive demonstration of dream manipulation by building on a number of previous discoveries about the way the brain works.

For instance, earlier studies had revealed that one class of brain cell acts like a curious biological GPS system: a particular "place cell" will burst into activity only when an animal enters a certain location in its environment – for example, a particular corner in a small arena that the animal is free to explore.

Neuroscientists also know exactly which parts of the brain light up with activity when a reward is on offer – a tasty morsel of food, for instance. Finally, it's also becoming clearer that the dreaming process has a role in processing the learning experiences that an individual has generated while awake.

Benchenane and his colleagues fitted their mice with tiny portable brain scanners and let them loose in a small arena. They took care to notice exactly which place cells in the rodent brain lit up when each mouse wandered into one particular corner of the arena.

Fast-forward a few hours and the mice have finished exploring the arena and have settled down for a nap. The scientists kept monitoring their brain activity as they entered the dreaming period of sleep – which

White mice had a false memory inserted into their brains.

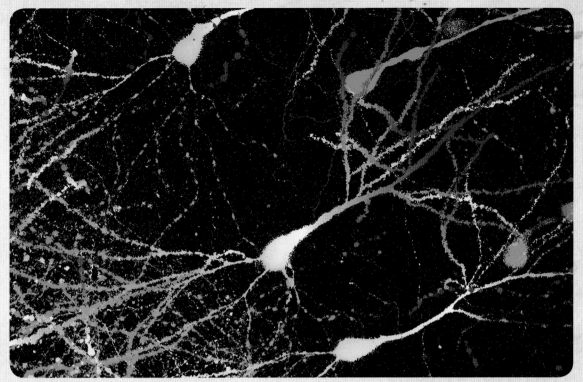

Mouse brain cells light up with activity during REM sleep.

all mammals and birds go through. In particular, the researchers were looking for the precise moment that the place cells associated with the arena corner lit up, suggesting that the mouse was revisiting that waking experience in its sleep.

At that very moment, the scientists manually stimulated reward centers elsewhere in the mouse brain in a bid to build a mental connection between the arena corner and the concept of a reward. They hoped that by doing so they could convince the mouse that it had received a reward when it visited that particular corner of the arena – which would perhaps change the animal's subsequent behavior.

It worked: upon awaking, the mice seemed far more attracted to the arena corner than they had been previously, suggesting that they did indeed have some expectation of receiving a reward there.

Studies like this show just how far science can now intrude into our personal thoughts and dreams. That might worry some – but there are real benefits to be had here, too. Post-traumatic stress disorder has been linked to malfunctions in the dreaming process: if scientists can learn how to "read" dreams as people sleep – and then step inside those dreams and modify them – they might be able to lessen the grip that a traumatic memory has on the person's waking life.

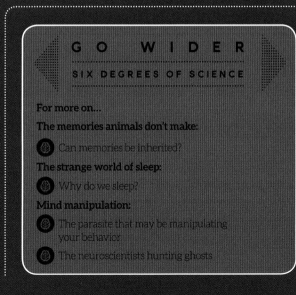

GO WIDER
SIX DEGREES OF SCIENCE

For more on...

The memories animals don't make:
- Can memories be inherited?

The strange world of sleep:
- Why do we sleep?

Mind manipulation:
- The parasite that may be manipulating your behavior
- The neuroscientists hunting ghosts

CAN BOREDOM BE FATAL?

It wasn't a big ask. College students in a 2014 experiment simply had to sit in a room alone and entertain themselves for 15 minutes.

To make the task more challenging, the students were denied the devices that many people rely on for personal entertainment. No mobile phone, computer or book. Not even pen and paper. In fact, the research team – led by Timothy Wilson at the University of Virginia, Charlottesville – offered the students only one possible means of distraction. They were given access to a button, and warned that pushing it would deliver a painful electric shock.

Two-thirds of the men in the experiment, and one-quarter of the women, chose to press the button. One man did so 190 times.

It's stark evidence of the extreme and often self-destructive efforts that people will go to in order to avoid boredom. Arguably, there are good reasons for doing so: studies suggest that both health and wealth can suffer when people grow bored. But are the risks so severe that someone can literally bore themselves to death?

Almost as long as people have been writing down their thoughts, there have been complaints of boredom. Some early civilizations, for instance, imagined their deities were prone to boredom. In one Ancient Egyptian myth, an omnipotent god creates the Earth partly to deal with the ennui of his immortal existence.

There's even the suggestion that sleep, long considered a mysterious process, might serve no other purpose than to kill time – and perhaps reduce the risk of boredom.

Avoiding boredom might well bring advantages. A 1990 study by Alex Blaszczynski at Westmead Hospital in Australia suggested pathological gamblers are often unusually prone to boredom. More recently, in 2005, Eric Dahlia's team at the University of Southern Mississippi in Hattiesburg asked college students to answer questions on their driving habits and their propensity to boredom. Those who claimed to be most easily bored also admitted to habits that are associated with dangerous driving.

It's perhaps understandable, then, that NASA and other space agencies worry about the risks boredom might pose if and when astronauts set out on the long and monotonous journey through space to Mars. But would those astronauts' lives actually be at risk purely because

Some people prefer physical discomfort to boredom.

(Above left) Boredom might
trigger creativity.

(Above right) Those who are addicted to
gambling may be more prone to boredom.

of the boredom they might experience?

It's difficult to establish a direct link between boredom and death. In 2009, however, two researchers at University College London in the UK – Annie Britton and Martin Shipley – decided to investigate.

They used information from questionnaires that some London-based civil servants are asked to complete every few years. As luck would have it, during the 1980s the questionnaire had asked the civil servants to rate their level of boredom.

Britton and Shipley found that many of the government employees who had claimed to be particularly bored 30 years ago are now dead. So are some of the employees who said they were never bored, of course, but a bit of number crunching showed that the death rate among the "always bored" workers is higher than it should be through chance alone. Cardiovascular disease, in particular, seemed to be a common cause of death among the bored.

The pair concluded that being bored can increase the likelihood of an early death – but almost certainly because boredom is just a sign of underlying risk factors. People who reported being particularly bored also tended to report being less physically active, for instance.

In fact, scientific attitudes to boredom are shifting. Many psychologists have begun linking boredom with creativity. Someone who has lost interest in the task at hand might let their mind wander. Their performance on the task will be poor, but their wandering mind might have come up with a completely different, and very creative, idea that is worth pursuing.

Today, some psychologists even encourage parents to allow their children to be bored from time to time, because doing so can increase their capacity for creative and independent thinking. Even so, as the electric shock experiment shows, it's probably wise not to push people too far into boredom.

GO WIDER

SIX DEGREES
OF SCIENCE

For more on...
The dangers of immortality:
 The secrets of
the immortal

The struggle to explain why we sleep
Why do we sleep?

The boring journey to Mars:
 Next stop: Mars

THE PEOPLE WITH A LARGE CHUNK OF BRAIN MISSING

A few years ago, a 24-year-old Chinese woman walked into a hospital complaining of dizziness, nausea and vomiting. Her doctors began performing routine tests and soon came up with a likely explanation for her symptoms. A large chunk of her brain was missing.

Where most people have a cerebellum, located at the back of the brain in the base of the skull, the Chinese woman had an empty cavity filled with fluid.

This case study, described by the woman's astonished doctors in 2014, is unusual. But it's not unique. In 2007, for instance, doctors in France reported that a man in his mid-40s had a large fluid-filled cavity where most of his brain should be.

These strange cases suggest that our brains are much more flexible and adaptable than scientists once thought possible.

The Chinese woman and French man are alike in more ways than one. Both clearly have very unusual brains. However, even more surprisingly, both were living reasonably normal and healthy lives before doctors realized significant areas of their brains were missing. Both had found partners and got married. The Chinese woman was a mother. The French man had fathered two children. At the time of his remarkable diagnosis he was holding down a standard job as a civil servant – he had only sought medical help because of a slight weakness in one leg.

Later his doctors established that, as a child, he had undergone a procedure to deal with water on the brain. It was perhaps through a rare and inadvertent side effect of this treatment that he had gradually lost brain tissue.

Exactly why the Chinese woman lacked a cerebellum is less clear. Her condition probably began early in her development, long before her birth.

As far as we know, both the Chinese woman and the French man continue to live healthy lives to this day,

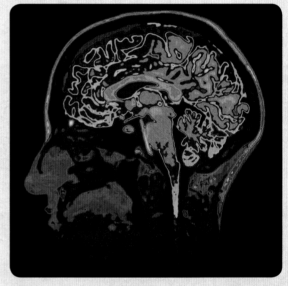

The human brain is built up of several regions, each thought to have its own speciality.

despite lacking so much brain tissue.

Discoveries like these seem to defy scientific expectations. The brain is built up of several regions, and each is traditionally imagined to be particularly suited for a specific task – like the specialized components of a computer. The cerebellum, for instance, helps us keep fine control over our muscles. It also has an important role to play in language. Meanwhile, two nearby parts of the brain called the hippocampi are apparently vital for memory and spatial navigation.

Some scientific discoveries reinforce the idea that different brain regions are highly specialized. London's taxi drivers must memorize the city's labyrinthine network of 25,000 streets to qualify for the job. Those who succeed in internalizing this map have larger hippocampi at the end of the process than they did when they began their training. Clearly, the hippocampi really do play a key role in memory.

But the rare reports of people missing parts of their brain – and experiencing only mild side effects as a consequence – suggest other brain regions must be able to take on new tasks if they have to. Evidently, neuroscientists still have plenty to learn about the human brain's potential for versatility and adaptability.

In fact, until 2013 it wasn't even completely clear that the human brain has any potential for change and renewal as we age. Jonas Frisén at the Karolinska Institute in Stockholm, Sweden, and his colleagues were the first to prove that it does. Using a dating method that is based on radioactivity levels in the atmosphere, Frisén's team could age individual cells in the brains of volunteers who had left their bodies to science. Their findings suggest that some regions of the brain gain new cells on a daily basis.

Individual brain regions clearly do have specializations, but perhaps they are not straitjacketed to behave in a certain way. Instead, certain parts of the brain may have the potential to improvise. They might be better viewed as jacks-of-all-trades and masters of one.

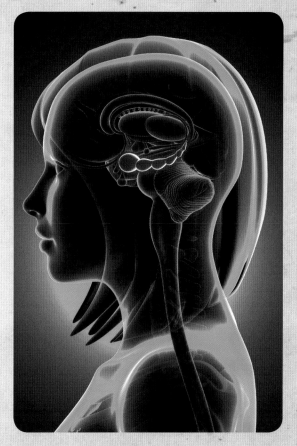

(Below left) Perhaps brain regions aren't quite like specialized computer components after all.

(Above) One of the brain's two hippocampi, which play a key role in memory.

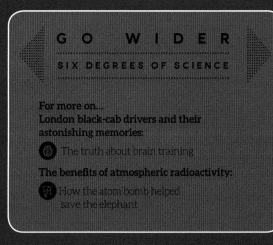

G O   W I D E R

SIX DEGREES OF SCIENCE

For more on...
London black-cab drivers and their astonishing memories:

The truth about brain training

The benefits of atmospheric radioactivity:

How the atom bomb helped save the elephant

CAN MEMORIES BE INHERITED?

Nothing else has quite the evocative power of our sense of smell. The scent of a certain flower, for instance, can take someone back to a specific event from their childhood. For one group of mice, though, a smell apparently took them back much further than their youth.

Astonishingly, it might have taken them all the way back to the moment, long before their birth, when their grandparents learned to associate the smell with an unpleasant experience.

There are plenty of characteristics we can inherit from our parents and grandparents. Our facial features, our height, even whether we are left or right handed – all are controlled to some degree by genes. These traits are passed from parent to child through DNA.

But there are also plenty of things we can't inherit through our genes. Learned experiences, in particular, don't travel through DNA. A parent might have picked up a dozen languages, worked out how to ride a bike, even learned through exposure how to fend off particular pathogens – but none of these lessons are coded into our

genes. Each generation must start from scratch and learn skills and experiences like these afresh.

Memories should be learned from scratch, too. But a controversial experiment from 2013 suggests that some of them might pass down the generations.

Neuroscientist Brian Dias at Emory University in Atlanta, Georgia, and his colleagues taught mice to associate a sweet scent with a small but unpleasant electric shock. Eventually, the mice cringed in anticipation of the shock at the mere whiff of the sweet scent.

Astonishingly, that association seemed to cross the generations.

The offspring of the mice also cringed in response to the sweet smell, although Dias and his colleagues had not actually taught these younger mice to associate the smell with the shock. Even the offspring of these offspring – the original rodents' grandchildren – cringed. Again, they did so without any formal training to link the sweet smell with an electric shock.

The findings aren't explained by some deep-seated loathing of the sweet smell shared by all mice. A "control" group of rodents that didn't undergo training to associate the scent with a shock never cringed at the sweet smell, and neither did their children or grandchildren.

Many other scientists were – and still are – deeply skeptical of Dias's findings. They point out that there just doesn't seem to be any obvious way that a memory, even a powerful and potentially important one, could alter an animal's DNA and pass down the generations.

That said, geneticists are still learning about the details of genetic inheritance. Most genomes carry huge quantities

There is nothing quite as evocative as our sense of smell.

of seemingly useless DNA – studying those regions might reveal new and unexpected ways in which genetic inheritance operates.

For instance, a few decades ago, no geneticist would have predicted that simple bacteria can encode learned experiences into their DNA and pass them down the generations. But recent research has confirmed that they can. If bacteria can pass experiences through their DNA, perhaps animals can, too. One study in 2016 identified one way it could be done in principle.

Our genomes can, in fact, be altered during our lives through our experiences. Small chemical molecules clamp onto particular sites in our DNA, making some genes more active, some genes less active, and switching off some genes altogether. However, the standard view is that these "epigenetic" changes are all reset in sperm and egg cells. Infants inherit a clean slate, with their DNA free from their parents' epigenetic influences.

However, biologist Jerome Jullien at the University of Cambridge and his colleagues challenged that standard view. They found that frog sperm cells might transfer some gene-altering chemical tags to offspring after all. Not only that, but these tags actually seem to be vital for the healthy development of frog tadpoles: for reasons that are still unclear, it might be important that epigenetic changes pass down the generations.

If gene-altering epigenetic tags can indeed jump between generations, it's just about conceivable that experiences and memories might be caught up in that process, too. Much more work must be done before that link can be made with confidence, but perhaps one day geneticists will be able to confirm that some of our memories were made long before we were born.

(Below left) Some mice seemed to "remember" events from their parents' lives.

(Above) DNA can be modified by life experiences.

GO WIDER
SIX DEGREES OF SCIENCE

For more on...
Making unusual memories:
Can science hack your sleeping mind?

The "useless" DNA in our genomes:
Is 90 percent of our DNA junk?

The bacteria that inherit learned experiences:
Gene editing just got serious

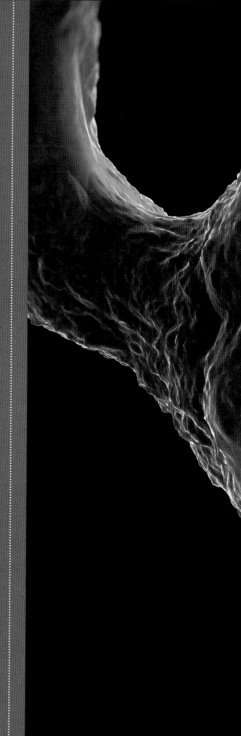

WHY
DO
WE
SLEEP?

For several hours each day, we are dead to the world. Our bodies enter a catatonic state, while our minds cross into a strange twilight zone where the rules of reality begin to break down. And scientists still don't really know why.

Sleep might help the brain keep the connections between brain cells working properly.

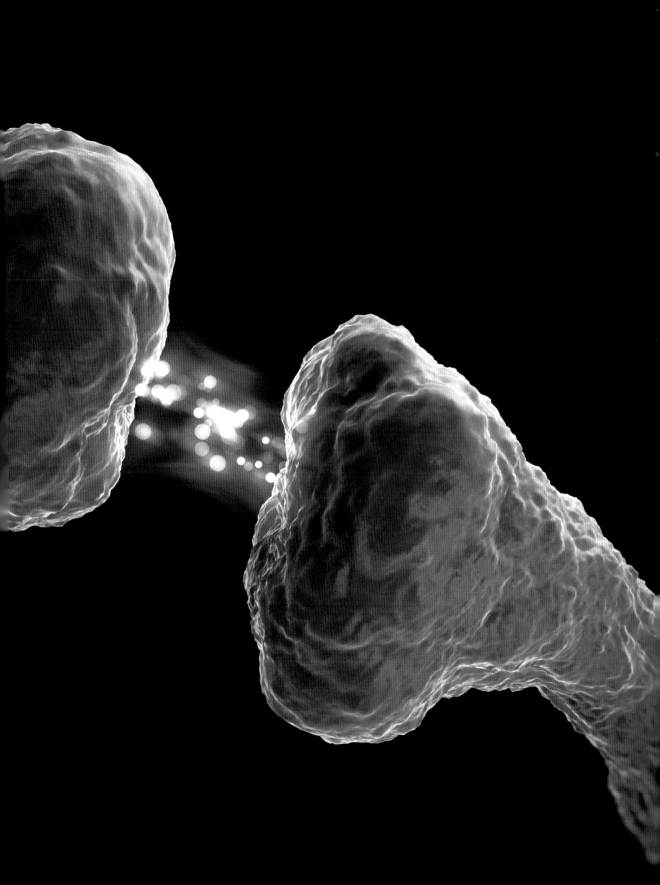

Sleep is one of the enduring mysteries of existence. Its purpose is the subject of endless debate. Some have suggested it's a way for animals to make small energy savings – body temperature can drop slightly when we doze. Others have speculated that sleep might simply offer a way to pass the time when an animal has nothing better to do.

But explanations like this don't quite ring true – particularly since studies in animals show that a lack of sleep proves deadly within days. Whatever purpose sleep serves, it's clearly vital for our health and well-being.

Some scientists – including Giulio Tononi of the University of Wisconsin-Madison – think they are close to solving the puzzle. They say sleep is all about providing the brain with some downtime for vital housekeeping.

In 2016, Tononi reported that mice lose some of the connections between brain cells as they sleep. He thinks sleep is about turning the day's experiences into memories, and consolidating those memories into a smaller number of connections, freeing up space in the brain so that the animal has room to record new experiences the following day.

Other scientists think brain housekeeping might involve different pathways. Our brain cells send messages to each other by releasing little streams of chemicals. Over time, though, these streams of "neurotransmitters" can build up into a sea of molecules that could clog the connection between two brain cells. Robert Cantor at Dartmouth College in Hanover, New Hampshire and some other neuroscientists think that sleep gives the brain an opportunity to flush out these neurotransmitters and keep the connection clear.

A remarkable discovery in 2012 supports the idea. A team led by Jeffrey Iliff, now at Oregon Health and Science University in Portland, discovered a network of vessels in the mammalian brain that no one had noticed before. This "glymphatic" network flushes out the fluids between brain cells – and a follow-up study in 2013 showed that it is most active during sleep.

Even if the "housekeeping" idea is correct, though, there is almost certainly more to sleep than this. Why, for instance, do we dream?

Dreaming is most commonly associated with the famous REM stage of sleep. The curious thing is

(Opposite) The mystery of dreams is every bit as puzzling as the mystery of sleep.

(Right) We dream for about one quarter of the time we are asleep.

that while almost all animals seem to sleep – even microscopic worms and tiny fruit flies – very few animals go through REM sleep. Humans and other mammals do, as do birds. In 2016 some researchers suggested reptiles might experience REM sleep, too. And that's about it.

A few neuroscientists think they know what dreaming animals have in common. Birds, mammals and some reptiles have relatively advanced brains and often live in complex social groups. To survive, they need to be able to learn from their experiences.

The trouble is, the events we experience are often distractingly visceral. They can be distressing, or embarrassing, for instance, which might make our brains reluctant to revisit them. One idea championed by researchers, including Matthew Walker at the University of California in Berkeley, is that dreams are the brain's way of carefully snipping away the emotion from an experience to leave just the raw, unemotional memory – which is then made easier to recall and refer to during the waking hours.

The idea might explain one unusual finding from 2015. David Samson at Duke University in Durham, North Carolina, realized that humans are in the dreaming REM stage for 25 percent of their time asleep. Other primates dream for less than 10 percent of their time asleep. This might be because human societies are so complex that we need to spend more time dreaming in order to process all of our daily experiences.

If dreaming is about converting emotional experiences into unemotional memories, distressing conditions like post-traumatic distress disorder (PTSD) might become a little easier to understand and perhaps treat. People with PTSD often experience vivid and emotionally crippling flashbacks. Perhaps that's because of a breakdown in their dreaming process, says Walker, which leaves them unable to strip away the emotion from experiences when they dream.

If Walker is right, finding ways to improve the sleeping and dreaming process might be one way to help people with PTSD. Dreams might actually be more powerful than many people realize.

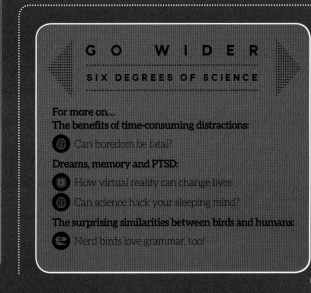

G O W I D E R

SIX DEGREES OF SCIENCE

For more on...
The benefits of time-consuming distractions:

🧠 Can boredom be fatal?

Dreams, memory and PTSD:

🧠 How virtual reality can change lives

🧠 Can science hack your sleeping mind?

The surprising similarities between birds and humans:

🐦 Nerd birds love grammar, too!

THE TRUTH ABOUT BRAIN TRAINING

Parkinson's disease is a devastating condition that can leave people with muscle stiffness and difficulty walking. Dementia can follow as the disease advances. There is no known cure, but one study from 2016 suggests there is a surprising way for some people with Parkinson's to manage their symptoms. They can help themselves. They just need to learn to control their thoughts.

The idea that someone – anyone – can train their brain and change their life has a long but checkered history. For much of the twentieth century, most neuroscientists thought brain training was very unlikely to work. The consensus view was that the brain becomes fixed by the time we reach adulthood: if it changes at all with age, it is only by degrading – just as it does with a condition like Parkinson's.

This consensus began to shift about 20 years ago, partly because of an ingenious set of experiments.

London black-cab drivers are required to memorize the city's extensive network of streets – called gaining "The Knowledge" – so that they can ferry a passenger to their destination along the most efficient route without consulting maps or a potentially unreliable GPS system.

As early as the year 2000, Eleanor Maguire at University College London, UK, and her colleagues had found evidence of something extraordinary. Brain scans showed that drivers who successfully internalized The Knowledge have unusually large posterior hippocampi – regions of the brain that neuroscientists know are particularly specialized for memory and spatial navigation.

But skeptics pointed out a flaw in the research: some people may naturally grow up with unusually large hippocampi, and perhaps this makes them more likely to succeed in learning The Knowledge. In other words,

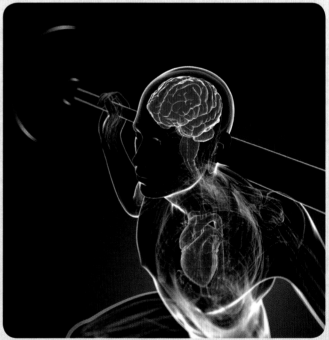

Many people today try to exercise mind as well as body.

perhaps brain training had nothing to do with the hippocampal growth.

A follow-up study in 2011 silenced those skeptics. Maguire and her colleagues studied a group of people from the moment they embarked on their quest to learn

(Above left) Learning to navigate London's streets changes the brains of taxi drivers.

(Above right) People can learn to control video games with their thoughts.

The Knowledge right up until the moment that some of them succeeded. All of the volunteers had hippocampi of a similar size at the start of the process. A few years later, those who had successfully completed the formidable memory task had larger hippocampi than those who had given up and pursued another career.

"Brain training" moved from the scientific margins to the mainstream – and confirmation, in 2013, that some parts of the brain gain new cells throughout life helped consolidate its position there.

According to many neuroscientists, brain training – when it's done the right way – does seem to make a real difference to people's lives. The people with Parkinson's who volunteered to take part in the 2016 study provide a great example.

A team led by David Linden at Cardiff University in the UK used a brain scanner to eavesdrop on the volunteers' brain activity as they moved the muscles in their hand. Later, the volunteers had to play a simple computer game by thinking about moving the muscles in their hand. As they played, Linden's team recorded their brain activity and fed it back into the game in real time, meaning that the more control the volunteers had over their thoughts the better they performed in the game.

The process left the volunteers better able to control their muscles and their disease symptoms outside the game environment.

Brain training is now being explored for its potential to help other conditions, including post-traumatic stress disorder. It has even become a vital component of a pioneering new protocol to help people who are paralyzed learn to walk again, with promising results.

It would be wrong to assume that the entire neuroscientific community has leapt on the brain training bandwagon, though. Many are still deeply uneasy about the idea. In 2014, a group of 70 leading researchers issued a statement questioning the benefits. In response, a second group of 120 scientists issued a statement of their own pointing to the benefits brain training can bring.

It's a controversy that could run and run.

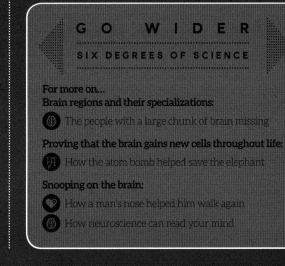

GO WIDER
SIX DEGREES OF SCIENCE

For more on...
Brain regions and their specializations:
The people with a large chunk of brain missing

Proving that the brain gains new cells throughout life:
How the atom bomb helped save the elephant

Snooping on the brain:
How a man's nose helped him walk again

How neuroscience can read your mind

THE WOMAN WITH THE SUPER VISION

The grass is iridescent. Vegetation is accented with delicate shades of violet. And, with their eye-catching rainbow of hues, the stones seem to pop out of the scenery. The world is a spectacular place when viewed through the eyes of Concetta Antico.

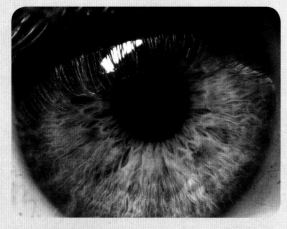

(Left) The world looks different through some people's eyes.

There's a reason why she sees such vibrancy in the landscape. She has a set of cells in her eyes that most people lack. With them, she can detect subtle shades that many of us miss.

Color is strange. It really exists only in our minds, which makes color far more subjective than we might imagine. The world's reaction to "the dress" illustrates this point very well. A photograph uploaded to the Internet in 2015 showed a dress that some viewers were convinced was black and blue – others were adamant it was white and gold.

Perhaps people, like Antico, with a condition called tetrachromacy could act as ultimate arbiters in such disputes.

Most humans are "trichromat." This means they carry three sets of light-sensitive "cone cells" in the retina at the back of each eyeball. One set is tuned very roughly to detect red light, one to detect green and one to detect blue – although all three can also pick up the colors around these three strongest shades, which is why most people have full color vision.

The genes for blue cone cells are located on one of the standard chromosomes in the human genome. We each carry two copies of these chromosomes. But by chance, the genes for red and green cone cells are located on the X sex chromosome. Women carry two copies of this chromosome – the XX condition – but men have just one copy coupled with the famous Y chromosome – the XY condition.

Genes naturally pick up mutations from time to time, and the "color" genes on the X sex chromosome seem to be particularly prone to doing so. If one of the red and green genes mutates, it can, in principle, end up tuned to a slightly different color – a darker red, for example.

This is what can lead to tetrachromacy.

A woman with tetrachromacy would carry genes for "dark red" cone cells on one of her X chromosomes and genes for normal non-mutated "red" cone cells on her second X chromosome – as well as genes for "green" cone cells and genes for "blue" cone cells. She would have eyes tuned to four different parts of the color spectrum.

Many women may be tetrachromat – in theory as many as one in every 10. Men can't develop the condition, though, because it requires the presence of two sex X chromosomes and men carry just one.

Proving which women actually are tetrachromats is very difficult. Antico is one of the few women to have had her tetrachromacy confirmed by scientists.

It was behavioral biologist Kimberly Jameson at the University of California, Irvine, and psychologist Alissa Winkler at the University of Nevada in Reno who helped demonstrate Antico's tetrachromacy. More recently, they have continued their research to try to get a sense of how the world must look through her eyes.

One of their studies was published in 2015. The scientists ran optical tests a little like those someone might encounter during a routine eye exam. By asking volunteers, including Antico, to compare subtly different shades, Jameson and Winkler could establish that Antico's extra color sensitivity is particularly active around the red end of the spectrum.

They could also establish that Antico sees colors vividly even in relatively gloomy conditions – conditions under which most people begin to lose the ability to distinguish between colors.

Even with this insight, it's as difficult to imagine what life must be like for Antico as it is for someone with normal vision to really appreciate what life must be like with color blindness. But as luck would have it, Antico is an artist, and when she works with oils her paintings have an unusual sense of vibrancy.

Studying her artwork may perhaps give mere trichromats some sense for the way the world can look through the eyes of someone with tetrachromatic vision.

(Above) Concetta Antico's paintings offer a glimpse into the landscape as she sees it.

The eye's cone cells (highlighted in green) are largely responsible for our color vision.

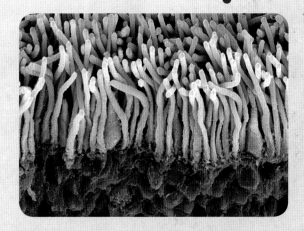

GO WIDER

SIX DEGREES OF SCIENCE

For more on...
The male-specific Y chromosome:

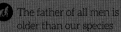 The father of all men is older than our species

Seeing the world through someone else's eyes:

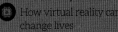 How virtual reality can change lives

THE NEUROSCIENTISTS HUNTING GHOSTS

Something spooky happens when neuroscientists and roboticists team up to perform experiments on willing human volunteers. They prove that ghosts are real.

Or, at least, they show how the human brain can be tricked into imagining that there is a ghostly presence nearby. Understanding how and why we sometimes conjure up ghosts is important work – it could help explain some features of brain conditions such as schizophrenia.

Almost one in five Americans claims to have seen a ghost.

Ghosts – or, more accurately, illusions of their presence – are surprisingly common. When Italian mountaineer Reinhold Messner wrote an account of his experiences at altitude, for instance, he recalled once having the uncanny feeling that a phantom mountaineer was climbing with him, always keeping just out of eyeshot.

Accounts like this fascinate neuroscientist Olaf Blanke at the Swiss Federal Institute of Technology in Lausanne. In 2014, Blanke and his colleagues wondered whether they could replicate such phantoms in the lab.

Some people seem to be particularly susceptible to experiencing the "ghost" illusion. Blanke and his colleagues tried to work out why. They scanned the brains of ghost-prone people and found that they often have slight damage to three distinct regions – called the temporoparietal cortex, the insular cortex and the frontoparietal cortex.

What links these three brain areas is that all are thought to help us assess the position of objects in the vicinity of our body.

This suggested to Blanke's team that they could trigger the ghost illusion in susceptible people by manipulating the objects surrounding their bodies. To do so, they came up with a strange experimental setup. Each volunteer was told to stand between two robot arm devices, facing one and with their back to the other. They were then asked to insert their "pointing" right index finger into a special hole in the "hand" of the robot arm in front. Finally, they were told to jab with their index finger. Doing so meant that the volunteers controlled the movement of the robot hand and arm in front.

What the volunteers didn't know was that the second robot arm – the one immediately behind them – was hooked up to the first arm in such a way that it instantly mirrored its movements. This meant that as a volunteer jabbed with their pointing finger, the second robot arm automatically poked them in the back.

The volunteers were asked to explain how they felt as they used this setup. Most reported the odd sensation that they were poking themselves in the back. But they didn't sense any ghostly presence.

Blanke's team then altered the setup slightly by introducing a delay to the system. Now, when the volunteers jabbed with their finger, they received a poke to their back a fraction of a second later. This small delay made a world of difference: when quizzed about their experience using this setup, some of the volunteers said they had the unnerving feeling that there was someone else standing nearby.

The delay programmed into the system apparently amplified the problems the volunteers already had with processing sensory information near their bodies. The result confirmed Blanke's hunch: ghosts materialize when someone's experience of the world immediately around them doesn't quite match their expectations.

It's not the only study that has seen neuroscientists step into the brain's Twilight Zone. Other researchers are as keen to explore reality-bending quirks like déjà vu. A leading theory is that we experience déjà vu when the brain constructs false memories out of half-remembered events. In 2016, Akira O'Connor at the University of St. Andrews, UK, and his colleagues came up with an alternative hypothesis.

They developed a technique that triggers déjà vu, and then scanned the brains of people at the very moment they were experiencing the weird phenomenon. They noticed that the process is triggered by areas of the brain

Robot arms can help conjure up a ghostly presence just out of eyeshot.

involved in decision making, not in memory. O'Connor thinks déjà vu occurs when the brain recognizes a conflict in its memories – essentially when the brain is confident it hasn't experienced something but still has the odd sensation that it actually has.

Work like this is fascinating for its own sake, but it might have practical purposes, too. People with conditions like schizophrenia sometimes have distressing hallucinations or delusions that they are losing control to some mysterious invisible presence. If research into phenomena like ghosts and déjà vu pinpoints where in the brain these problems begin, science may move a step closer to understanding life with schizophrenia and developing more effective treatments.

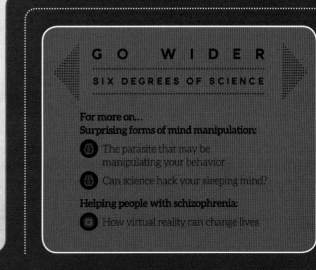

GO WIDER

SIX DEGREES OF SCIENCE

For more on...
Surprising forms of mind manipulation:

- The parasite that may be manipulating your behavior
- Can science hack your sleeping mind?

Helping people with schizophrenia:

- How virtual reality can change lives

CHAPTER EIGHT

HUMANITY'S PAST, PRESENT & FUTURE

EXTINCT HUMANS FOUND IN OUR DNA

From the thin air of the high Tibetan Plateau to the stifling humidity of tropical forests to the frigid bleakness of the Arctic, our species has learned to survive and prosper in an extraordinary range of different environments. There's no doubt that we did so partly through our ingenuity and our almost endless capacity for creativity. But that's not the full story. We had some help from Neanderthals and other long-extinct humans.

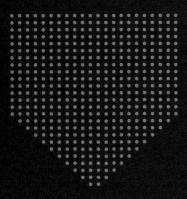

Neanderthals are extinct but their genetic legacy lives on.

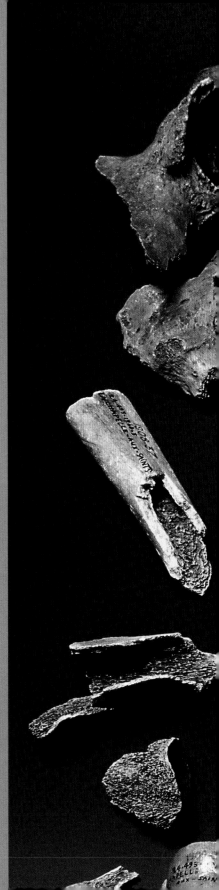

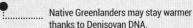
Native Greenlanders may stay warmer thanks to Denisovan DNA.

Our human species – *Homo sapiens* – first appeared in Africa about 315,000 years ago. About 100,000 years later, humans began trickling out of Africa – first into Asia and then on until they reached Australia, Europe and eventually the Americas.

As our species roamed across the world, it encountered all sorts of other species of human, whose ancestors had left Africa much earlier in prehistory.

The most famous of these "other" humans were the Neanderthals, who lived in Europe and west Asia. However, there were also strange "Hobbits," standing just 1 meter (3.2 feet) tall, living in parts of Indonesia – and one still-mysterious group of humans that lived in east Asia. Scientists still know very little about this particular group. They have been dubbed the "Denisovans."

There were almost certainly even more ancient humans living in Eurasia, too. Tentative signs of one emerged in a genetic analysis published in 2016.

Our ancestors didn't simply eye these other humans from a distance. They interacted with them – and they occasionally had children with some of them.

We know this because of revolutionary advances in genetics technology. Scientists can now extract DNA from the bones of some extinct humans and reconstruct their ancient genomes.

Starting in 2010, geneticists began to look at those ancient genetic sequences in detail, and they realized that some parts of the Neanderthal and Denisovan genomes perfectly match sequences in living humans. By far the most likely explanation for the match is that DNA from these ancient humans entered our gene pool in the Stone Age.

Since 2011, scientists have begun to find signs that this ancient DNA played a crucial role in helping our species adapt to life outside Africa.

One study led by Peter Parham at the Stanford School of Medicine in California focused on the immune system. Neanderthals had spent thousands of years adapting to deal with the diseases that occur in Europe and Asia. Our species hadn't, because we evolved in Africa – but the offspring of sex between our human species and Neanderthals had the potential to inherit Neanderthal DNA that makes fighting some Eurasian diseases a little easier. Over time, those offspring spread the useful DNA through the modern human gene pool, meaning our

G O   W I D E R
SIX DEGREES OF SCIENCE

Tibetans can cope better with thin high-altitude air, perhaps because of Denisovan DNA.

species gained the ability to fight off Eurasian diseases just as well as Neanderthals could.

The mysterious Denisovans helped our species, too. In 2014, Emilia Huerta-Sanchez at the University of California in Merced and her colleagues realized that most Tibetans carry Denisovan DNA in part of a gene that helps them breathe the thin air of the Tibetan Plateau. It's possible that the Denisovans had adapted to live at altitude and, when they bred with some members of our species, passed on the useful adaptation.

By 2015, Huerta-Sanchez and her colleagues had found evidence of another Denisovan signature – this time in indigenous communities living in Greenland. The Denisovan DNA seems to help these people produce more brown fat, an unusual form of body fat that generates heat in cold conditions. Again, Denisovans might have evolved adaptations like this themselves to deal with life in freezing environments, and passed them to members of our species.

Joshua Akey at the University of Washington in Seattle calls these chunks of DNA "genomic gifts."

Not all of the genomic gifts were beneficial. Some Neanderthal DNA seems to make modern humans more prone to conditions such as type 2 diabetes, liver disease and hay fever.

In retrospect, this genomic gift giving didn't do Neanderthals and Denisovans much good, of course. It helped our species grow stronger and invade deeper into new territory – their territory. Perhaps this forced Neanderthals, Denisovans and other human species like the Hobbits to marginal environments where it was tougher to survive.

By about 30,000 years ago, all of those other species of human had vanished. Our species had conquered the world.

For more on...
Revolutionary DNA studies:

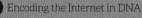 Encoding the Internet in DNA

The father of all men is older than our species

Dramatic events from prehistory:

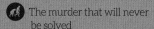 The murder that will never be solved

The flood story that isn't a myth

The surprising science of brown fat:

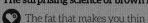 The fat that makes you thin

THE FATHER OF ALL MEN IS OLDER THAN OUR SPECIES

DNA ancestry tests have become a great way for people to find out a little more about their family history. But in 2013, one family in the US probably got more than they bargained for when they submitted a sample for analysis. The man it belonged to had a unique genetic heritage stretching back about 340,000 years – long before our species evolved in Africa.

His DNA led to an astonishing and paradoxical conclusion. The father of all men lived and died more than 25,000 years before our species even evolved.

The defining genetic male feature is the Y chromosome, a chunk of the genome that men carry and women lack. The Y chromosome makes up about two percent of the male human genome, and it is passed exclusively between fathers and their sons.

This feature makes the Y chromosome special. Imagine a man is studying his family tree, and goes back three generations to his eight great grandparents. Our man's genome contains a random mix of DNA from these eight people – but his Y chromosome comes from just one of them: his father's father's father. In fact, our man has just one "Y chromosome ancestor" in every generation. His Y chromosome came from just one of his 16 great great grandparents, from just one of his 32 great great great grandparents, and so on.

In reality, the Y chromosome a man inherits from his father isn't necessarily an identical copy. The Y chromosome (and DNA in general) picks up subtle genetic changes at a roughly steady rate – say, one every few hundred years.

It's handy for geneticists that it does. Those genetic changes give the scientists a way to estimate when two people last shared a common ancestor just by measuring how closely their DNA matches: two people with a shared grandfather will have more in common at the DNA sequence level than two people with completely different ancestry.

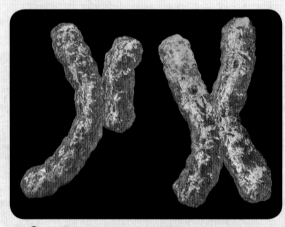

The Y chromosome next to an X chromosome.

But there is only so much genetic variation in human DNA. Geneticists have been collecting DNA samples from volunteers all across the world: every single Y chromosome sequence they have examined is, fundamentally, rather similar. Using what they know about the rate DNA accumulates changes, the geneticists can calculate that all men alive shared a Y chromosome ancestor about 140,000 years ago.

Given that the Y chromosome passes exclusively between fathers and their sons, this leads to a surprising conclusion. Every man on the planet shares the same father's father's father's (and so on) father – an unknown man who was alive about 140,000 years ago, almost certainly in Africa.

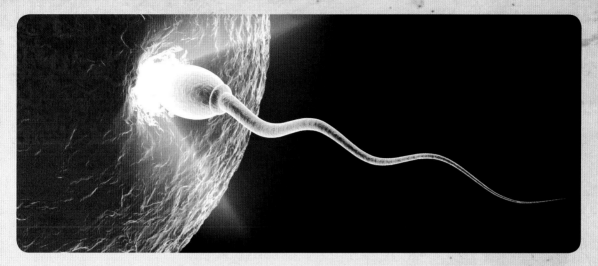

Or so geneticists thought until 2013. Then, during a routine DNA ancestry test in the US, a team led by Michael Hammer at the University of Arizona in Tucson found a Y chromosome that didn't fit this global picture.

At the genetic level, it had a sequence strikingly different from any Y chromosome previously analyzed. It came from an American man, Albert Perry, of relatively recent west African descent – and Perry couldn't possibly have descended from the "father of all men" who lived 140,000 years ago.

The scientists calculated that it would have taken about 340,000 years, not 140,000, for the unusual Y chromosome to accumulate so many genetic changes. In light of the astonishing discovery, the real "father of all men" – the man who gave his Y chromosome to all living men including the few who carry Perry's unusual version – must have lived at this much earlier point in prehistory. He lived long before our species even evolved.

Surprisingly, there is an explanation for this.

Our species might have appeared only 315,000 years ago, but plenty of ancient human species predated us.

 The Y chromosome passes down the generations each time a father has a son.

An important point to bear in mind is that these ancient humans didn't vanish as soon as our species evolved: they actually lived alongside our species for tens of thousands of years before they finally went extinct. During that long period of co-existence, our species and these ancient human species occasionally interbred.

Perhaps a few thousand years ago, in west Africa, there lived a man belonging to one of those ancient human species – a man who had a Y chromosome that was strikingly different from any seen in a member of our human species.

If this man had a sexual encounter with a female of our species, who then gave birth to a son, the strange Y chromosome will have jumped into our modern gene pool. That brief sexual encounter had a lasting legacy: as a consequence, geneticists now have to dig surprisingly deep into prehistory to find the father of all men.

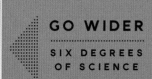 **GO WIDER**

SIX DEGREES
OF SCIENCE

For more on...
The ancient humans that came before our species:

 Extinct humans found in our DNA

 The murder that will never be solved

The disadvantages of carrying the Y chromosome:

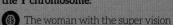 The woman with the super vision

The surprises lurking in human DNA:

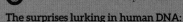 Is 90 percent of our DNA junk?

THE SECRETS HIDDEN INSIDE YOUR BODY

How well do you know your own body? In November 2016, J. Calvin Coffey and D. Peter O'Leary at the University of Limerick in Ireland realized they weren't quite as familiar with human anatomy as they thought. They argued that an obscure set of tissues in the abdomen form one continuous structure that deserves to be recognized as a distinct human organ. If other scientists agree, the "mesentery" would become the 79th organ in the human body.

The find is a reminder that there are still secrets hidden inside the human body, even after centuries of intense scientific scrutiny.

Scientists are probing the human body using ever more sophisticated equipment. Perhaps it's inevitable that they will occasionally discover something brand new.

In 2013, for example, a team of surgeons led by Steven Claes at University Hospitals Leuven in Belgium discovered a ligament in the knee – the anterolateral ligament – that had somehow failed to be specifically identified in almost all previous detailed scientific studies. The ligament seems to play an important role in knee stability – some people treated for anterior cruciate ligament tears don't regain full control of their knee, apparently because of damage to the anterolateral ligament.

The human brain is no less mysterious. In 2012, a team led by Jeffrey Iliff, now at Oregon Health and Science University in Portland, found a completely new system of vessels running through the mouse brain. The "glymphatic system" – which has now been seen in the human brain, too – seems to be a bespoke waste disposal system to flush unwanted chemical junk out of the brain. This includes amyloid-beta, a substance that accumulates in the brain of people with Alzheimer's disease, which suggests the glymphatic system might hold clues for treating the condition.

Geneticists have been making new and unexpected discoveries about the human body, too. In 2001, as they

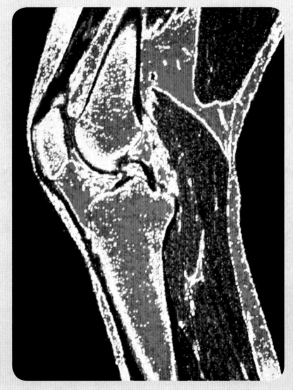

The knee contains a ligament that wasn't formally recognized until 2013.

(Opposite above) Some people have feet that bend slightly in the middle when they walk.

began to analyze the human genome in detail, they discovered that our bodies are built using just 20,000 or so genes – a fraction of the number that most geneticists were expecting. Those genes collectively comprise just two percent of the human genome, leading to serious questions about what, if anything, the other 98 percent of the DNA in our cells is doing.

Anatomical discoveries are not restricted to the abdomen and knee. The bones of our "midfoot" – the area between the toes and the heel – are pulled into a solid and rigid structure by ligaments. This ancient adaptation makes it easier for us to walk on two legs – chimps have a flexible midfoot instead, which is one reason why they find walking on two legs awkward.

But in 2013, Jeremy DeSilva and Simone Gill, both then at Boston University in Massachusetts, filmed people walking barefoot and realized as many as one in 13 of us actually have bendy chimp-like feet without realizing it. No one knows why, although the fact that people now routinely wear rigid-soled shoes might be a factor. Our shoes might mean that our feet no longer need to maintain a rigid midfoot to help us walk efficiently.

One of the biggest and most surprising discoveries of the last 15 years is that we are never truly alone. Each human body is home to roughly 40 trillion microbes, many living in our gut. Doctors have now discovered they can replace "bad" gut microbes with "beneficial" ones. That's another surprise, and one that has led to the rise of fecal transplants – a new procedure where gut microbes in the feces of healthy people are transplanted into the gut of the sick.

Even something as simple as the human nose has the capacity to surprise: some scientists have found new and potentially invaluable antibiotics from studying the bacteria that live there.

In other words, anyone with ambitions to make their mark in science and discover something brand new could do a lot worse than to look inwards.

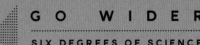

GO WIDER

SIX DEGREES OF SCIENCE

For more on...

The mysteries locked in our DNA:

- What synthetic life reveals about the living world
- The genes that light up after death

The human microbiome:

- What lives in the fourth domain?
- The life-saving power of excrement

The astonishing human nose:

- Crunch time for antibiotics
- Are human beings quantum machines?

THE MURDER
THAT WILL
NEVER BE SOLVED

It was a vicious and frenzied attack. The right-handed assailant dealt the young victim several blows to the forehead. Two were so severe that they punctured the skull and penetrated the brain. Then, the attacker threw the body down a deep rocky chasm – perhaps in an effort to conceal the crime.

But even the long arm of the law can't reach far enough to bring this murderer to justice. The crime was committed about 430,000 years ago. It is the earliest recorded murder – and it predates the appearance of our species by hundreds of thousands of years.

Scientists who study human origins often have little more than a few scraps of fossil bone and teeth to work with. The researchers working at a Stone Age site called Sima de los Huesos in the Atapuerca Mountains of northern Spain are far more fortunate. In a single deep chasm they have found remains belonging to no fewer than 28 ancient humans, all members of an extinct human species that was closely related to the famous Neanderthals.

One leading theory is that the chasm served as some sort of primitive tomb – a sacred place where the ancient humans disposed of their dead. This idea would suggest that each of the 28 individuals was cast into the pit after death, perhaps as part of some sort of communal and cathartic ritual.

But the remains of one of the 28 individuals don't fit this poignant picture.

This individual is best represented by a skull, which the scientists reconstructed from 52 fragments that they collected from the chasm. The skull belonged to a young adult of uncertain gender. And it carries two large holes in its forehead.

The holes looked highly suspicious. Nohemi Sala from the Complutense University of Madrid, Spain, and her

One skull from Sima de los Huesos has two large holes in its forehead.

colleagues decided to investigate further with state-of-the-art forensic technology.

They generated a high-resolution 3D computer model of the skull to get a better sense of how the holes might have formed. The precise way the bone has fractured around each hole is consistent with the idea that both predate death – dry bone cracks in a distinctly different way.

The holes themselves, meanwhile, are very similar to each other in size and shape. Both were almost certainly inflicted using the same blunt object, which appears to

Ancient stone tool or potential murder weapon?

(Right) Sima de los Huesos is traditionally interpreted as a sacred prehistoric site for the disposal of the dead.

have been rotated slightly between the two blows. The size and shape of the holes suggests the weapon was a spear or stone hand axe – both weapons of choice for Stone Age humans.

Using evidence like this, the scientists gradually worked towards a disturbing conclusion. They realized there was a good chance they were looking at an ancient murder victim, attacked from the front by someone armed with a lethal Stone Age weapon.

The exact identity of the assailant will never be known. But there's a very good chance that they were right handed. Both injuries are on the left side of the skull, consistent with the idea that the assailant was wielding the murder weapon in their right hand.

It's possible – although extremely difficult to prove – that the murderer then tried to conceal their crime by disposing of the body in a chasm chosen by locals as a place to lay their dead.

Work like this is a reminder that forensic science is not just about modern crime scene investigation. The tools that scientists have developed can give us a fascinating, if sometimes unsettling, glimpse into what life and death was like for the ancient humans who lived long before our species evolved.

GO WIDER

SIX DEGREES OF SCIENCE

For more on...
Breakthroughs in forensic science:

How the atom bomb helped save the elephant

The ancient humans who predate our species:

Extinct humans found in our DNA

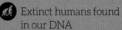 The father of all men is older than our species

WAS THIS HUMANITY'S BIGGEST MISTAKE?

In 2006, archaeologists made an extraordinary discovery in an ancient burial ground. They were investigating the skeletons associated with a 9,000-year-old early farming community in Pakistan. Many of the prehistoric humans had dental cavities – and, remarkably, nine seemed to have seen a prehistoric dentist to have some of the rotten tissue carefully drilled away. The findings revealed that dentistry predates civilization by thousands of years.

Early farming communities settled down in one spot and built the first villages.

There's another reason why the discovery is significant. It fits with a controversial idea: that farming was the worst mistake humanity ever made.

Humans became serious about farming their food about 10,000 years ago. The decision changed everything. Before agriculture, humans generally lived in small bands, moving across the landscape to hunt wild animals and gather seeds, fruits, nuts and tubers – whatever was available and nutritious.

Farming ended this way of life. People became tied to a particular location to tend their crops and raise their livestock: villages appeared. Farmers generated so much food that there was plenty to feed many mouths throughout the year. Populations grew, and some members of the society were freed from the daily toil of

Childbirth might have been made more painful and risky by farming.

food production to develop new skills – making pots and working metals, for instance.

Humans had hunted and gathered food for tens of thousands of years without drastically changing their lifestyles. A few thousand years after the first crops were planted, people were living in cities, developing writing and living socially complex lives. No wonder it's called the farming revolution.

However, US scientist and author Jared Diamond thinks we focus too much on the benefits farming brought. As long ago as the 1980s, he was arguing that farming was the "worst mistake in human history." He says farming made life much tougher for early humans.

Farming brought war. There was certainly violence – even murder – long before farming. But if two hunter-gatherer groups fell out, they could easily walk away from each other to solve the problem. Farmers were tied to the land – their land. If they fell out with a neighboring group, they had to stay and fight. It's notable that evidence of death on a large scale doesn't appear in the archaeological record until after the agricultural revolution.

Farming brought inequality. Although some anthropologists disagree, many see hunter-gatherer groups as egalitarian. Farming may have generated food surpluses, but that also allowed some people to gain power by controlling those supplies. Social elites appeared, and the lower classes soon became unhappy. Some of the earliest texts, written more than 4,000 years ago in the Mesopotamian city-state of Lagash, are bitter complaints about a corrupt ruling class.

And farming brought disease. Large populations squeezed together in towns were a breeding ground for transmissible infections. The impact of health can be seen in ancient human skeletons. Most obviously, early farmers were significantly shorter than their hunter-gatherer predecessors.

But a close look into ancient jaws reveals something else: farming brought widespread tooth decay. This is almost certainly a consequence of a dietary change. Hunter-gatherers ate a protein-rich diet, whereas farmers adopted a diet rich in carbohydrates including sugar. Dental health was generally good before farming. Afterwards, cavities – and dentists – became an unfortunate fact of life.

The farming revolution was so dramatic and so recent, relatively speaking, that humans might still be feeling its shockwaves and adapting to its effects. Many women today find childbirth a difficult and painful process. One idea is that it only became so after the arrival of farming, because a developing baby puts on more pre-birth weight if the mother is eating a carbohydrate-rich diet. Consequently, it's a tighter squeeze when the baby passes through the birth canal.

For all of these disadvantages, the world as we recognize it today depends on farming – particularly with the global population still rising. This means it's vital that science continues to explore ways to boost crop yields to try to make sure there is still enough food to feed everyone. Farming might have been the worst mistake in history, but a modern world without farmers is unthinkable.

G O   W I D E R

SIX DEGREES OF SCIENCE

For more on...
The problems associated with farming:

Are we living through the Anthropocene?

Crunch time for antibiotics

The future of farming:

The future of the hamburger

When food bites back

How beer could help cure malaria

Humanity's struggle to adapt to farming:

The power of the placebo

Why our brains are shrinking

IS 90 PERCENT OF OUR DNA JUNK?

There's a popular urban myth about the human brain. We each use just 10 percent of our brain's full potential – and if only we could somehow tap into the other 90 percent, we could become a population of Einsteins.

It's not true. But perhaps it's not that surprising that the myth endures, given that some equally astonishing scientific statements are true. For instance, it's true that all of the stars and planets account for just 4 percent of the known Universe – the remaining 96 percent remains frustratingly difficult to study.

What, then, should we make of another statement making the rounds: that as much as 90 percent of the human genome is little more than worthless junk? Is this concept of "junk" DNA another urban myth, or is it true?

The human genome, which is written in DNA stored inside most of our cells, is extraordinarily large. So large, in fact, that if you were to take the DNA from just one human cell and carefully stretch it out, the DNA strand would measure about 2 meters (6.5 feet) in length. It's little wonder that it took geneticists several years to "read" the three-billion-long sequence of genetic letters encoded in that DNA molecule.

They completed the job in the year 2000 and then turned to the task of analyzing the genome. By 2001, they had come to a surprising conclusion: the human genome contains just 20,000 to 30,000 genes – not the 100,000 that many geneticists were expecting. Collectively, the DNA that "codes" for those genes comprises just two percent of the genome.

What is the other 98 percent doing?

Some of it turns out to be doing quite a lot. Geneticists have begun studying the 98 percent of the genome that doesn't code for genes – sometimes called "non-coding DNA" – and they have made an unexpected discovery. Non-coding DNA has played an important role in

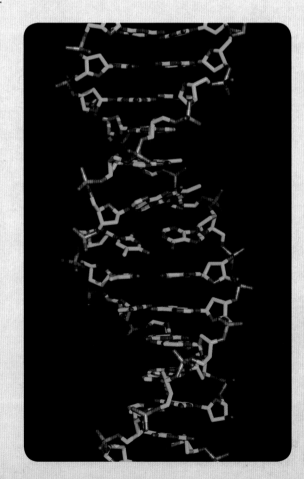

The human genome is written in DNA's ⋯⋯⋯⋯● famous double helix.

Many human features may have been shaped by DNA once labelled as "junk." ....●

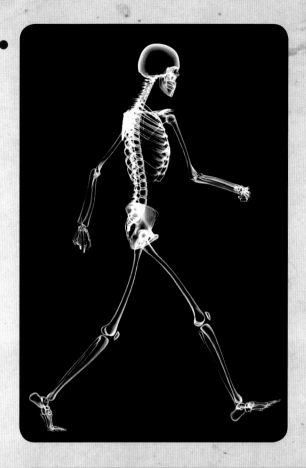

evolution in general, and human evolution in particular. In 2008, for instance, a team led by James Noonan at Yale University in New Haven, Connecticut, compared stretches of non-coding DNA in the human and chimp genomes. Our ancestral line split from the chimp line about six or seven million years ago, and over that time, the human and chimp genomes have been evolving independently at a steady rate.

But Noonan's team found that some stretches of non-coding DNA in the human genome have evolved remarkably quickly since we split from chimps, which usually occurs only in DNA that is performing an important function. The scientists' studies suggest that one of these stretches of non-coding DNA plays an executive role when our hands and feet are developing in the womb: it controls whether certain genes are switched on or off.

Noonan and his colleagues speculate that this piece of non-coding DNA might even help orchestrate the development of an iconic human trait: our opposable thumbs.

More recently, stretches of non-coding DNA have been linked to other important human traits – our ease with walking upright on two legs, our large brains and even our ability to control our face muscles when we talk.

Clearly, such important stretches of DNA can hardly be considered "junk," even if they don't actually code for genes. But all of the important stretches of non-coding DNA so far identified add up to somewhere between 5 and 10 percent of the human genome. That means about 90 percent of the genome still has no known purpose and seems to deserve "junk" status.

Some geneticists suspect that most – maybe even all – of that "junk" will eventually turn out to serve some useful purpose as they continue to explore the human genome.

But many evolutionary biologists are not so sure. They have identified ways in which the human genome can gradually accumulate worthless genetic junk – and they say the "cost" of carrying around this junk is so low that evolution hasn't bothered to tidy it up and throw it away. These evolutionary biologists suspect that the geneticists will one day concede that 50 percent or more of the human genome really is just junk.

Until the debate is settled one way or the other, it's impossible to be sure how much junk DNA our genomes really carry.

GO WIDER

SIX DEGREES OF SCIENCE

For more on...
Mysteries of the Universe:

The mystery at the core of the Universe

Unexpected insights into the evolutionary process:

The walking fish that learned fast

Can memories be inherited?

The father of all men is older than our species

The technological benefits of DNA research:

Encoding the Internet in DNA

THE FLOOD STORY THAT ISN'T A MYTH

In those olden days, there was a large plain extending from the main land out to the White-topped Rocks, about nine miles out from Cape Chatham. On one occasion, two women went far out on the plain, digging roots ... After a while they looked up, and saw the sea rushing towards them over the great plain.

Robert Matthews, a self-taught anthropologist, heard this story in the early days of the twentieth century from an aboriginal group living on Australia's southwest coast. What Matthews didn't know was that the story might document a real flooding event – one that hit the area at least 7,000 years ago.

Antarctica lost a significant amount of ice 12,500 years ago, triggering sea level rise.

The flooding story might be something truly exceptional: an account of an event that predates written history.

Real-life catastrophes or other types of dramatic events can often end up woven into local oral history. Over time, though, stories become embellished – new elements are added, others are removed. Most scientists assume that 500 to 800 years after the event actually occurred, the story will have drifted so far from the original narrative that it will have lost all of its meaning.

Some of the stories told by the aboriginal groups around Australia's coast seem to have escaped this fate. Here, oral folk stories may have retained accuracy for thousands of years.

Two Australian-based researchers – Patrick Nunn at the University of the Sunshine Coast in Queensland and Nicholas Reid at the University of New England in New South Wales – concluded as much in 2015 after a careful examination of the evidence.

The pair hunted through old literature for traditional folk stories told in 21 different aboriginal communities scattered all around Australia's long coastline. Every single community had at least one story on a remarkably similar theme: a narrative that referred back to a time when the sea rose and permanently inundated an area of previously dry and usable land.

 Australia's coastline has been stable for the last 7,000 years.

For such a similar story to be part of the oral tradition in so many independent communities, some separated from each other by thousands of kilometers, Nunn and Reid say we must at least consider the possibility that they all refer to a real event. All of the stories may describe a dramatic time when the sea level around Australia rose permanently by several meters.

Yet scientific study of Australia's coastline tells us this sort of sea level rise hasn't happened for thousands of years. After the peak of the last Ice Age, about 24,000 years ago, glaciers and ice sheets slowly began to retreat and shrink. As the meltwater flowed into the ocean, sea levels began to rise.

There was a particularly abrupt period of sea level rise about 12,500 years ago, at roughly the same time that the ice sheet over Antarctica lost a great deal of mass. Perhaps it is this phase of sea level rise to which the stories refer. Certainly, by about 7,000 years ago, sea levels had reached heights we would recognize today, and they haven't changed much since.

If the Australian stories really do date back to those prehistoric times, though, how have they survived for so long without losing accuracy?

 Modern technology may be more of a draw than ancient stories.

The answer, according to Nunn and Reid, might lie in the highly formalized way that stories are passed down from generation to generation. The communities have developed a series of crosschecks – young storytellers are tested by others in the social group to make sure they have the details of the story correct.

It may also help that Australia's aboriginal people were isolated from the rest of humanity for thousands of years. No large invading populations arrived to interfere with or dilute the narratives, until Europeans reached Australia a few centuries ago.

With unique factors like this at play, it's possible that the narratives have survived down 300 or 400 generations, according to Nunn and Reid.

But all things must pass. The two researchers told reporters that they fear these ancient stories might soon be told for the last time. For perhaps the first time in thousands of years, younger members of these coastal communities are no longer particularly interested in their cultural heritage. Dunn and Reid say youngsters find modern technology – computers, smart phones and the like – far more alluring.

If the researchers are correct, stories that have been communicated faithfully down hundreds of generations might soon be silenced forever because of the distraction of today's global communication gadgets.

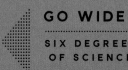

GO WIDER

SIX DEGREES OF SCIENCE

For more on...
Humanity's prehistory:

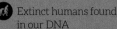 Extinct humans found in our DNA

The rise of smart phones:

 The end of the audio jack

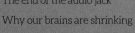 Why our brains are shrinking

THE POWER OF THE PLACEBO

They've been called miracle pills: tablets that seem to help people cope with conditions as diverse as cancer, Parkinson's disease and heart disease. Best of all, they cost next to nothing.

That's because these pills contain no actual medicine. They are simple sugar pills – placebos – and they seem to work because the person taking them has an expectation that they will help. However, if our bodies can fight off some diseases all by themselves, why do they only do so after we have taken a placebo?

Some scientists think they have an answer – and it's tied up with the idea that humans haven't fully evolved to twenty-first-century living.

One of the most curious things about the placebo effect is that it's not a simple side effect of modern medicine. Animals show a placebo response, too.

In 2002, Randy Nelson at Ohio State University in Columbus and his colleagues injected hamsters with a substance to mimic a bacterial infection. The hamsters didn't try to fight off the "infection" if a timer tuned to match the short days of winter controlled when the lights above their cage turned on and off. But if the cage lights were switched on for longer to mimic summer, the hamsters mounted a strong immune response. The cage lights seem to be acting as a weird placebo.

By 2012, a team led by Pete Trimmer, then at the University of Bristol in the UK, had tested an explanation for this sort of placebo effect – one that links the effect to the way our immune system works.

The basic idea is very simple: the immune system sucks up a great chunk of an animal's available energy resources. On a subconscious level animals – including humans – are reacting to signals in their environment to "decide" whether or not to ramp up an immune response.

For hamsters, it makes sense to avoid mounting a full-on immune response in midwinter. There's not much food around, and cranking up the immune system could drain resources from other body systems and leave the hamster at a significantly greater risk of dying. It's better to suffer through a mild illness, mounting only a minimal immune response.

In summer, the situation is different. There's plenty of food, and the hamster can "afford" a big immune response.

Sugar pills are actually playing a similar role in humans, thinks Trimmer. Subconsciously, people's bodies may decide not to mount a large immune response against an illness because of the chance that the immune system

Russian hamsters show an unusual placebo response.

could then rob other bodily systems of the resources they need in order to function properly. A sugar pill masquerading as real medicine is a bit like the hamster's summer, he says: because people believe the pills will weaken their illness and make it easier to fight, their body subconsciously "decides" to mount a full immune response after the placebo is swallowed.

When Trimmer's team explored this idea with a computer simulation, they made an extra prediction: if someone's faith in the placebo is too great, the body might subconsciously decide that it doesn't need to fight a disease at all – it might think that the placebo will destroy the disease all on its own. In these people, the immune system might switch off altogether and their condition might worsen after they take the sugar pill. It's an idea some scientists called the "anti-placebo effect" – or a "nocebo" effect.

There's one obvious flaw when it comes to applying these ideas to humans, though. Many people eat a healthy, energy-rich diet on every single day of the year. They could "afford" to mount a full immune response at any time – but their bodies still apparently work in an austerity mode, assuming that a proper immune response risks becoming a dangerous drain on resources.

Nicholas Humphrey at the London School of Economics, UK, thinks that might just be because our bodies haven't caught up with the reality of twenty-first-

Grain stores should have helped reduce the placebo effect in humans.

century living. Our bodies still "think" we live through seasons of feast and famine – even though that hasn't really been the case since people switched from hunting and gathering their food to farming crops and livestock.

If and when our bodies do get used to the idea that they don't have to keep such careful control over our immune systems, maybe the placebo effect will vanish.

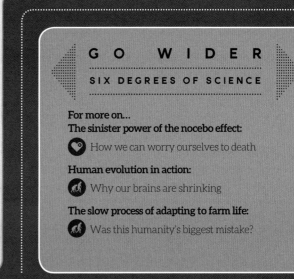

GO WIDER

SIX DEGREES OF SCIENCE

For more on...
The sinister power of the nocebo effect:
How we can worry ourselves to death

Human evolution in action:
Why our brains are shrinking

The slow process of adapting to farm life:
Was this humanity's biggest mistake?

ARE HUMAN BEINGS QUANTUM MACHINES?

Just over a century ago, a great German chemist – Alfred Stock – began synthesizing a brand new class of chemical: the boranes. It was dangerous work, given that boranes can be very explosive. But Stock was brave enough to give his new chemicals a sniff. He later wrote that they had a *ganz widerlich geruch* – a "very disgusting smell."

Ninety years later, Stock's words would provide some of the best evidence for a very controversial idea about our noses. According to one scientist, we depend on the weird effects of quantum physics to pick up a new scent on the air.

....

The human nose has an astonishing ability to distinguish between odors. Until recently, scientists thought it could easily discriminate between an impressive 10,000 smells. A 2014 study, led by Leslie Vosshall at the Rockefeller University in New York City, suggests that figure is actually hopelessly conservative: the scientists concluded that the human nose might detect one trillion different odors.

There's a standard explanation for how the nose does this. Each odor molecule has a slightly different shape. When one of these molecules "docks" onto a smell receptor in our nose, the receptor assesses the shape and identifies the molecule to work out its scent.

But Luca Turin, a biologist at the BSRC Alexander Fleming Institute in Vari, Greece, doesn't buy this idea. He thinks our noses actually use quantum physics to distinguish between smells.

Chemical molecules don't just have a distinct shape, they also have a distinct vibrational frequency, because of the way chemical bonds in molecules shake. It's this vibrational characteristic that Turin thinks helps

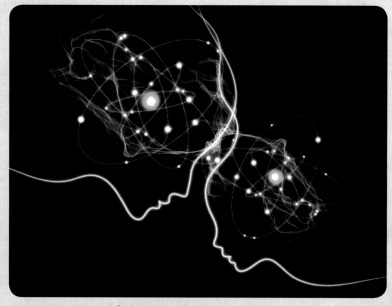

At a fundamental level our bodies might obey the strange rules of quantum physics.

us detect a scent. When an odor molecule comes into contact with a smell receptor in our noses, providing the molecule vibrates at just the right frequency, an electric signal can pass across the receptor and travel up to the brain, registering the odor.

The apparent flaw in Turin's idea is that the electrical

signal – in the form of tiny particles called electrons – shouldn't actually be able to pass across the receptor. This receptor is basically a very small but insurmountable barrier, and electrons – think of them as tiny tennis balls – don't have enough energy to smash their way through.

Turin, however, points out that this is only a flaw if physicists think of electrons as behaving according to the standard rules of "classical" physics. But electrons are so small that they can actually obey the strange and unintuitive rules of "quantum" physics.

One of these strange rules is that a tiny object like an electron can behave as a tennis-ball-like particle while simultaneously behaving as a wave of energy. The precise properties of that wave of energy – including its exact position in space – are essentially impossible for physicists to define accurately. They can estimate the exact location of an electron, but it's really just a well-informed guess: the position is surrounded by a halo of uncertainty, like a thick atmosphere surrounding a planet. In principle, the electron could actually be anywhere in that halo.

When the electron comes up against the barrier in the receptor, the halo of uncertainty is large enough that it extends across the barrier and incorporates atoms on the far side of the receptor. Given that the electron can theoretically be anywhere in the halo, it can actually now be on the other side of the barrier – even though that seems absurd.

The chances that the electron really has done the impossible and just popped up on the far side of the barrier are boosted by the presence of an odor molecule. If it's vibrating at just the right frequency, the electron can dump a little bit of its energy onto the molecule and attune itself precisely to the atoms beyond the barrier. Now it is more likely that the electron has indeed just popped up on the far side and that its signal can reach the brain.

Turin thinks that this apparently nonsensical process – which physicists call "quantum tunneling" – is going on in our noses with every sniff we take.

If he is right, all molecules that vibrate in the same way will smell the same, even if their shape is different. If he is wrong, and smell is actually about molecular shape, as biologists have long assumed, then molecules will only smell the same if they have the same shape.

To test the idea, Turin looked at the sulfurous molecules that make rotten eggs stink. Then he pored through chemical textbooks to find a class of molecule that vibrates in exactly the same way as those sulfurous molecules, despite having a completely different shape.

Eventually he found what he was looking for: boranes. And, going back to Stock's initial work on the boranes 90 years ago confirmed that they do indeed smell foul – just like rotten eggs. It was a finding that helps bolster Luca's ideas.

Even so, plenty of scientists remain skeptical about Turin's "quantum nose" claim. He will need to catch smell receptors in the very process of exploiting quantum tunneling if he is to win over more doubters. As of early 2018, he was optimistic about his chances of success.

The smell of rotten eggs might come down to the way odor molecules vibrate.

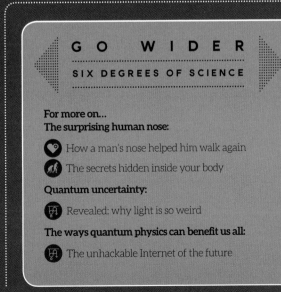

GO WIDER
SIX DEGREES OF SCIENCE

For more on...
The surprising human nose:

How a man's nose helped him walk again

The secrets hidden inside your body

Quantum uncertainty:

Revealed: why light is so weird

The ways quantum physics can benefit us all:

The unhackable Internet of the future

WHY OUR BRAINS ARE SHRINKING

We've come a long way. It's more than 13.8 billion years since the Big Bang; about 4.54 billion years since Earth formed; a little over half a billion years since animals evolved; some two million years since the first ancient humans emerged; and 315,000 years since our particular species appeared.

....

And the journey isn't over yet. Genetic studies under discussion at a scientific meeting in May 2016 emphasize that humans are evolving in front of our very eyes. For instance, ever since the Romans left Britain a few thousand years ago, the British seem to have become more tolerant of the lactose in milk, and more likely to be blue-eyed and blond-haired. There's also evidence that human lifespans are lengthening – and medical advances might push them up even further.

However, these features might seem relatively small compared to another change that is sweeping through the global human population. Our brains are shrinking.

It's easy to imagine that humans are now smarter than they have ever been – after all, our ancestors spent millions of years evolving larger and more sophisticated brains.

But brain size seems to have peaked about 20,000 years ago. On average, human brains are smaller in our current high-tech age – the Anthropocene as it is sometimes dubbed – than they were during the Stone Age.

John Hawks, an anthropologist at the University of Wisconsin–Madison who studies human evolution, says the difference is larger than you might expect: our brains have lost a volume about the size of a tennis ball.

The unanswered question is: why?

In 2011, Marta Lahr at the University of Cambridge, UK, pinned the blame on farming. A diet based on farming produce is just not as nutritious as the wild game, berries and nuts diet that hunter-gatherers enjoyed before the rise of agriculture, she says.

What's more, farming has encouraged people to live

Stone tools may have helped human brains grow.

cheek-by-jowl: disease spreads easily, and so more of the energy that humans now consume has to go into fending off infection, not growing large brains.

But there are plenty of other explanations for the brain drain. Some researchers say humans have just evolved leaner, sleeker brains that are "wired" more efficiently. Emerging neuroscientific evidence is compatible with this idea: studies involving people who have undertaken extraordinary memory challenges, for instance, show the brain has an under-appreciated ability to adapt and change, even within a lifetime.

There is another, more depressing explanation, though. Our species might be gradually losing its intellectual capacities. Biologist Gerald Crabtree at Stanford University, California, made this suggestion in 2012. Our ancestors needed to be quick witted with razor-sharp minds to survive. In the modern world, most people

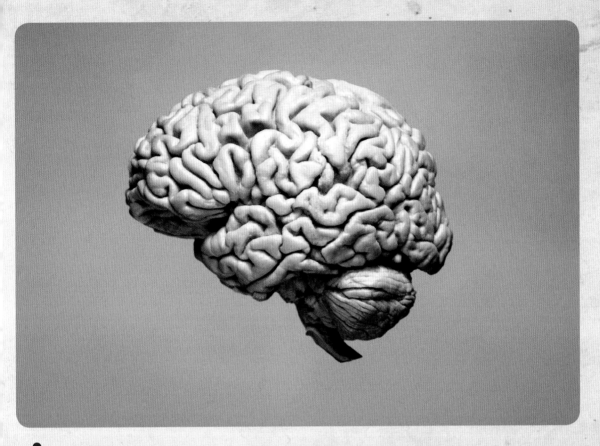

Our Ice Age ancestors had larger brains than we do today.

don't face the same daily struggle for existence. We can survive even if our brains are performing sub-optimally.

It's even been argued that the very technology that seems to mark us out as super intelligent is actually accelerating our intellectual decline.

A 2016 study by Benjamin Storm at the University of California, Santa Cruz, and his colleagues found that the smart phones many people carry have become expensive external brains. Because they connect us to the Internet – a vast repository of all sorts of knowledge – people are losing the need to remember important information, or even the need to work out how to solve problems on their own. The Internet tells us what to do.

About 3.3 million years ago, our ancient ancestors picked up stone tools for the first time and dramatically broadened their intellectual horizons. Today, the state-of-the-art tools many of us use every day might be narrowing ours.

SIX DEGREES OF SCIENCE

For more on…

Extending the human lifespan:

♥ Is this the secret to eternal youth?

The surprisingly devastating impact of farming:

🐗 Was this humanity's biggest mistake?

🐒 The power of the placebo

Smart phones:

▣ The end of the audio jack

🐒 The flood story that isn't a myth

BIBLIOGRAPHY

SPACE

The first stars in the Universe
Astrophysics: primordial stars brought to light. (30 September 2015)
Nature
Chronology of the Universe. (n.d.) *Wikipedia.*

The end of everything
Is Earth in danger of being hit with a gamma-ray burst? (23 March 2013) *Futurism*
On the role of GRBs on life extinction in the Universe.
(13 November 2014) *arXiv*
Did a gamma-ray burst initiate the late Ordovician mass extinction?
(5 August 2004) *International Journal of Astrobiology*

The extraordinary story of Planet Nine
Caltech researchers find evidence of a real ninth planet.
(20 January 2016) *Caltech*
Planet Nine hunters enlist big bang telescopes and Saturn probe.
(24 February 2016) *New Scientist*

The tiny spacecraft with big ambitions
Stephen Hawking and Yuri Milner launch $100m star voyage. (12 April
2016) *The Guardian*
Breakthrough Starshot. (n.d.) *Wikipedia*
Discovery of potentially Earth-like planet Proxima b raises hopes for life.
(24 August 2016) *The Guardian*

How to get rich in space
$195 billion in metal and fuel will just fly past the Earth.
(14 February 2013) *mining.com*
Asteroid miners can learn a lot from Philae's bumpy landing.
(30 July 2015) *Wired*
Asteroid-mining plan would bake water out of bagged-up space rocks.
(18 September 2015) *space.com*

Why there could be a parallel Universe
Ultimate guide to the multiverse. (23 November 2011) *New Scientist*
Horizon problem. (n.d.) *Wikipedia*

The weirdest star in the galaxy
The most mysterious star in our galaxy. (13 October 2015) *The Atlantic*
"Alien megastructure" star may be explained by interstellar junk. (19
September 2016) *New Scientist*

Next stop: Mars
Can humans hibernate in space? (27 April 2015) *The Guardian*
"Mars mission" crew emerges from yearlong simulation in Hawaii.
(29 August 2016) *npr.org*

The waves that distort our planet
The detection of gravitational waves was a scientific breakthrough, but
what's next? (April 2016) *Smithsonian Magazine*
What will gravitational waves tell us about the Universe?
(17 February 2016) *New Scientist*

The worlds beyond our solar system
Space oddities: 8 of the strangest exoplanets. (15 August 2013)
Popular Mechanics

Diamond planet worth $26.9 nonillion. (12 October 2012) *forbes.com*

PHYSICS

Why Earth's core is younger than you think
Earth's core is two-and-a-half years younger than its crust.
(22 April 2016) *New Scientist*
Real-world relativity: the GPS navigation system. (28 October 2016)
Ohio State University

Does particle physics have a problem?
The particle that wasn't. (5 August 2016) *The New York Times*
Physics crunch: Higgs smashes into a dead end. (27 February 2013)
New Scientist

How to hide information in the fabric of time
How to cloak a crime in a beam of light. (16 November 2010)
New Scientist
Time cloak used to hide messages in laser light. (28 November 2014)
New Scientist

Have we broken the light barrier?
Faster than light? CERN findings bewilder scientists. (23 September 2011)
Los Angeles Times
Flaws found in faster-than-light neutrino measurement.
(22 February 2012) *Nature*

How the atom bomb helped save the elephant
Cold War bomb testing is solving biology's biggest mysteries.
(11 December 2013) *pbs.org*
Carbon from nuclear tests could help fight poachers. (1 July 2013)
bbc.co.uk

The magma that could kill us all
Earth's time bombs may have killed the dinosaurs. (27 July 2011)
New Scientist
An ancient recipe for flood-basalt genesis. (18 August 2011) *Nature*

The mystery at the core of the Universe
Morphing neutrinos provide clue to antimatter mystery.
(12 August 2016) *Nature*

The unhackable Internet of the future
Chinese satellite is one giant step for the quantum Internet.
(27 July 2016) *Nature*
Why quantum satellites will make it harder for states to snoop.
(24 August 2016) *New Scientist*

Revealed: why light is so weird
Simultaneous observation of the quantization and the interference
pattern of a plasmonic near-field. (2 March 2015)
Nature Communications

One-third of the Earth is named after this man
Lucky strike in search for Earth's most common mineral.
(27 November 2014) *New Scientist*
Mineral kingdom has co-evolved with life, scientists find.
(14 November 2008) *Science Daily*

⬛ TECHNOLOGY

The real-life invisibility cloaks
How do you make a building invisible to an earthquake? (September 2012) *Smithsonian Magazine*
How could we build an invisibility cloak to hide Earth from an alien civilization? (14 April 2016) *The Conversation*

How virtual reality can change lives
Virtual reality took me inside the mind of a schizophrenic. (16 February 2015) *The Daily Dot*
Healing minds with virtual reality. (2 April 2015) *pbs.org*
The effect of embodied experiences on self-other merging, attitude, and helping behavior. (15 Feb 2013) *Media Psychology*

The future of the hamburger
Team wants to sell lab-grown meat in five years. (15 October 2015) *bbc.co.uk*
Lab-grown meat is in your future, and it may be healthier than the real stuff. (2 May 2016) *The Washington Post*

The end of the audio jack
Apple says it took "courage" to remove the headphone jack on the iPhone 7. (7 September 2016) *The Verge*
The 19th-century plug that's still being used. (11 January 2016) *bbc.co.uk*

Dinosaur resurrection: meet the chickenosaurus
Dawn of the chickenosaurus. (12 March 2016) *inquisitr.com*
From chicken to dinosaur: Scientists experimentally "reverse evolution" of perching toe. (22 May 2015) *Science Daily*

How to build a star on Earth
Nuclear fusion, the clean power that will take decades to master. (17 May 2015) *The Conversation*
Chinese fusion test reportedly reaches new milestone. (15 February 2016) *phys.org*
Wendelstein 7-X fusion device produces its first hydrogen plasma. (3 February 2016) *Max Planck Institute of Plasma Physics*

What synthetic life reveals about the living world
"Minimal" cell raises stakes in race to harness synthetic life. (24 March 2016) *Nature*
Artificial cell designed in lab reveals genes essential to life. (24 March 2016) *New Scientist*

The drones that control the weather
DRI unmanned cloud-seeding project gains ejectable flare capability. (23 June 2016) *Desert Research Institute*
Does cloud seeding work? (19 February 2009) *Scientific American*

Let your car do the driving
Volvo dashboard sensors take aim at drowsy driving. (19 March 2014) *automotive-fleet.com*
Self-driving Tesla was involved in fatal crash, U.S. says. (30 June 2016) *The New York Times*

Encoding the internet in DNA
The first book to be encoded in DNA. (20 August 2012) *Time*
Communicating with aliens through DNA. (18 August 2012) *Scientific American*

⬛ ENVIRONMENT

Are we living through the Anthropocene?
The Anthropocene: a new epoch of geological time? (31 January 2011) *Philosophical Transactions of the Royal Society A*
Scientists Say a New Geological Epoch Called the Anthropocene Is Here. (29 August 2016) *Time*

Saving the planet one eruption at a time
Geoengineering the planet: first experiments take shape. (26 November 2014) *New Scientist*
Dumping iron at sea does sink carbon. (18 July 2012) *Nature*

When food bites back
RNAi: The Insecticide of the Future. (23 May 2016) *University of Maryland*
The go-between: Life's unexpected messenger. (10 September 2014) *New Scientist*

The dry country that could water the world
Israel Proves the Desalination Era Is Here. (29 July 2016) *Scientific American*

Is a 250-million-year-old extinction event killing humans today?
Chinese coal formed during Earth's greatest extinction is still deadly. (1 July 2010) *Wired*
Coal combustion and lung cancer risk in XuanWei: a possible role of silica? (July 2011) *La Medicina del lavoro*

Has Chernobyl become a haven for wildlife?
Wolves, boar and other wildlife defy contamination to make a comeback at Chernobyl. (5 October 2015) *The Conversation*
At Chernobyl and Fukushima, radioactivity has seriously harmed wildlife. (25 April 2016) *The Conversation*

The killers lurking in Earth's ice
Biggest-ever virus revived from Stone Age permafrost. (5 March 2014) *New Scientist*
Methane release from melting permafrost could trigger dangerous global warming. (13 October 2015) *The Guardian*

The truth about green energy
The dystopian lake filled by the world's tech lust. (2 April 2015) *BBC Future*
A Scarcity of Rare Metals Is Hindering Green Technologies. (18 November 2013) *Yale Environment 360*

Can super-coral save our seas?
Ruth Gates' research to reverse rapid coral reef decline supported by Paul G. Allen. (4 August 2015) *University of Hawai'i News*
Unnatural selection. (18 April 2016) *New Yorker*

When life kills itself
Peter Ward: a theory of Earth's mass extinctions. (February 2008) *TED*
Paleontologist Peter Ward's "Medea hypothesis": Life is out to get you. (13 January 2010) *Scientific American*

🌐 NATURAL WORLD

The microbes that eat (and poo) electricity
Live wires: The electric superorganism under your feet.
(15 December 2010). *New Scientist*
Seabed superorganism uses electricity to lock up greenhouse gas.
(21 October 2015) *New Scientist*

Gene editing just got serious
5 Big Mysteries about CRISPR's Origins. (12 January 2017)
Scientific American
Swedish Scientist Begins Editing Human DNA in Healthy Embryos.
(25 September 2016) *Futurism*

Why cancer is like a selfish animal
Tumours could be the ancestors of animals. (9 March 2011) *New Scientist*
The Self-ish Cell: Cancer's emerging evolutionary paradigm.
(9 August 2011) *EvMed Review*

What lives in the fourth domain?
Glimpses of the Fourth Domain? (18 March 2011) *Discover*
Mystery microbes in our gut could be a whole new form of life.
(11 November 2015) *New Scientist*

The dwarf dinosaurs of Transylvania
The world's largest dinosaur can be seen right now in New York City.
(15 January 2016) *The Verge*
Did dinosaurs exist as dwarfs? (7 September 2014) *BBC Earth*

The walking fish that learned fast
Scientists raised these fish to walk on land. (27 August 2014) *The Verge*
Adapt first, mutate later: Is evolution out of order? (14 January 2015)
New Scientist

The secrets of the immortal
Greenland shark may live 400 years, smashing longevity record.
(11 August 2016) *Science*
The animals and plants that can live forever. (19 June 2015) *BBC Earth*

Could a frozen squirrel help humans cheat death?
When your veins fill with ice. (11 March 2016) *BBC Earth*
What the Supercool Arctic Ground Squirrel Teaches Us about the Brain's
Resilience. (26 June 2012) *Scientific American*

When superfast shrimps attack
How some animals accelerate faster than all others.
(19 September 2016) *BBC Earth*
The Most Powerful Movements in Biology. (September 2015)
American Scientist

Nerd birds love grammar, too!
Humans and birds share the same singing genes. (11 December 2014)
New Scientist
These birds use a linguistic rule thought to be unique to humans. (8
March 2016) *Washington Post*

♥ HEALTH & WELL-BEING

Crunch time for antibiotics
Keep medicine out of the dark ages. (2 July 2014) *Financial Times*
Antibiotic-Resistant Bacteria Are No Match For Medieval Potion.
(30 March 2015) *Popular Science*

The genes that light up after death
Hundreds of genes seen sparking to life two days after death.
(21 June 2016) *New Scientist*

How we can worry ourselves to death
The nocebo effect: how we worry ourselves sick. (29 March 2013)
New Yorker
New Insights into the Placebo and Nocebo Responses.
(31 July 2008) *Neuron*

The fat that makes you thin
Brown Fat, Triggered by Cold or Exercise, May Yield a Key to Weight
Control. (24 January 2012) *The New York Times*
Supercharging Brown Fat to Battle Obesity. (15 July 2014)
Scientific American

The life-saving power of excrement
Taboo transplant: How new poo defeats superbugs. (15 December 2010)
New Scientist
Policy: How to regulate fecal transplants. (19 February 2014) *Nature*

Human organs on demand
Anthony Atala: Printing a human kidney. (March 2011) *TED*
Soon, Your Doctor Could Print a Human Organ on Demand. (May 2015)
Smithsonian Magazine

How beer could help cure malaria
Pharma to fork: How we'll swallow synthetic biology.
(9 April 2014) *New Scientist*
Synthetic biology's first malaria drug meets market resistance.
(23 February 2016) *Nature*

Is this the secret to eternal youth?
Anti-ageing pill pushed as bona fide drug. (17 June 2015) *Nature*
Feature: The man who wants to beat back aging.
(16 September 2015) *Science*

How a man's nose helped him walk again
UCL research helps paralysed man to recover function. (21 October 2014)
University College London
The paralysed man who can ride a bike. (4 March 2016) *bbc.co.uk*

The first human head transplant
The Audacious Plan to Save This Man's Life by Transplanting His Head.
(September 2016) *The Atlantic*
Head transplant carried out on monkey, claims maverick surgeon.
(19 January 2016) *New Scientist*

🧠 THE BRAIN & HUMAN BEHAVIOR

The parasite that may be manipulating your behavior
How Your Cat Is Making You Crazy. (March 2012) *The Atlantic*
Cat parasite linked to mental illness, schizophrenia.
(5 June 2015) *CBS News*

How neuroscience can read your mind
Scientists use brain imaging to reveal the movies in our mind.
(22 September 2011) *Berkeley News*
Voicegrams transform brain activity into words.
(31 January 2012) *Nature*

Can science hack your sleeping mind?
False memories implanted into the brains of sleeping mice.
(9 March 2015) *The Guardian*

Can boredom be fatal?
Psychology: why boredom is bad... and good for you.
(22 December 2014) *BBC Future*
Bored to death? (21 December 2009) *International Journal of Epidemiology*

The people with a large chunk of brain missing
A civil servant missing most of his brain challenges our most basic theories of consciousness. (2 July 2016) *Quartz Magazine*
How Much of the Brain Can You Live Without? (n.d.) *Brain Decoder*

Can memories be inherited?
Fearful memories haunt mouse descendants. (1 December 2013) *Nature*
First evidence that sperm epigenetics affect the next generation.
(13 April 2016) *New Scientist*

Why do we sleep?
What is the real reason we sleep? (18 March 2016) *BBC Earth*
Mystery of what sleep does to our brains may finally be solved.
(12 July 2016) *New Scientist*

The truth about brain training
The brain's miracle superpowers of self-improvement.
(24 November 2015) *BBC Future*
Cache Cab: Taxi Drivers' Brains Grow to Navigate London's Streets.
(8 December 2011) *Scientific American*

The woman with the super vision
The mystery of tetrachromacy: if 12% of women have four cone types in their eyes, why do so few of them actually see more colors?
(17 December 2015) *The Neurosphere*
The woman with superhuman vision. (5 September 2014) *BBC Future*

The neuroscientists hunting ghosts
Ever felt a ghostly presence? Now we know why. (12 November 2014)
New Scientist
Mystery of déjà vu explained – it's how we check our memories.
(16 August 2016) *New Scientist*

🦍 HUMANITY'S PAST, PRESENT & FUTURE

Extinct humans found in our DNA
The 4 genetic traits that helped humans conquer the world.
(20 April 2016) *New Scientist*
DNA analysis reveals how humans interbred with Neanderthals.
(18 March 2016) *Wired*

The father of all men is older than our species
The father of all men is 340,000 years old. (6 March 2013) *New Scientist*

The secrets hidden inside your body
Mesentery: New organ discovered inside human body by scientists
(and now there are 79 of them). (3 January 2017) *The Independent*
1 in 13 people have bendy chimp-like feet. (29 May 2013) *New Scientist*

The murder that will never be solved
Scientists discover 430,000-year-old murder in Spain. (27 May 2015)
Popular Archaeology

Was this humanity's biggest mistake?
How our ancestors drilled rotten teeth. (29 February 2016) *BBC Earth*
The real reason why childbirth is so painful and dangerous.
(22 December 2016) *BBC Earth*

Is 90 percent of our DNA junk?
You are junk: Why it's not your genes that make you human.
(27 July 2016) *New Scientist*
How much of human DNA is doing something? (5 August 2014)
Genetic Literacy Project

The flood story that isn't a myth
Ancient Aboriginal stories preserve history of a rise in sea level.
(12 January 2015) *The Conversation*
The Atlantis-style myths that turned out to be true. (19 January 2016)
BBC Earth

The power of the placebo
Evolution could explain the placebo effect. (6 September 2012)
New Scientist
How a dog's mind can easily be controlled. (18 October 2016) *BBC Earth*

Are human beings quantum machines?
A quantum sense of smell. (24 March 2015) *Physics World*
Human nose can detect 1 trillion odours. (20 March 2014) *Nature*

Why our brains are shrinking
If Modern Humans Are So Smart, Why Are Our Brains Shrinking?
(20 January 2011) *Discover*
Are Humans Becoming Less Intelligent? (12 November 2012) *Live Science*

INDEX

PICTURE CREDITS